Radical Reorganization of Existing Work Structures Through Digitalization

Punita Duhan
Meera Bai Institute of Technology, India

Komal Singh
Meera Bai Institute of Technology, India

Rahul Verma
Bhai Parmanand Institute of Business Studies, India

A volume in the Advances in Knowledge Acquisition, Transfer, and Management (AKATM) Book Series

IGI Global
DISSEMINATOR OF KNOWLEDGE

Published in the United States of America by
IGI Global
Business Science Reference (an imprint of IGI Global)
701 E. Chocolate Avenue
Hershey PA, USA 17033
Tel: 717-533-8845
Fax: 717-533-8661
E-mail: cust@igi-global.com
Web site: http://www.igi-global.com

Library of Congress Cataloging-in-Publication Data

Names: Duhan, Punita, 1973- editor. | Singh, Komal, 1988- editor. | Verma, Rahul, 1989- editor.
Title: Radical reorganization of existing work structures through digitalization / Punita Duhan, Komal Singh, and Rahul Verma, editors.
Description: Hershey, PA : Business Science Reference, [2018]
Identifiers: LCCN 2017015916| ISBN 9781522531913 (hardcover) | ISBN 9781522531920 (ebook)
Subjects: LCSH: Information technology--Economic aspects. | Technological innovations--Economic aspects. | Digital communications. | Business enterprises--Technological innovations.
Classification: LCC HC79.I55 R33 2018 | DDC 338/.064--dc23 LC record available at https://lccn.loc.gov/2017015916

This book is published in the IGI Global book series Advances in Knowledge Acquisition, Transfer, and Management (AKATM) (ISSN: 2326-7607; eISSN: 2326-7615)

British Cataloguing in Publication Data
A Cataloguing in Publication record for this book is available from the British Library.

Advances in Knowledge Acquisition, Transfer, and Management (AKATM) Book Series

ISSN:2326-7607
EISSN:2326-7615

Editor-in-Chief: Murray E. Jennex, San Diego State University, USA

MISSION

Organizations and businesses continue to utilize knowledge management practices in order to streamline processes and procedures. The emergence of web technologies has provided new methods of information usage and knowledge sharing.

The **Advances in Knowledge Acquisition, Transfer, and Management (AKATM) Book Series** brings together research on emerging technologies and their effect on information systems as well as the knowledge society. **AKATM** will provide researchers, students, practitioners, and industry leaders with research highlights surrounding the knowledge management discipline, including technology support issues and knowledge representation.

COVERAGE

- Cognitive Theories
- Cultural Impacts
- Information and Communication Systems
- Knowledge acquisition and transfer processes
- Knowledge management strategy
- Knowledge Sharing
- Organizational learning
- Organizational Memory
- Small and Medium Enterprises
- Virtual communities

IGI Global is currently accepting manuscripts for publication within this series. To submit a proposal for a volume in this series, please contact our Acquisition Editors at Acquisitions@igi-global.com or visit: http://www.igi-global.com/publish/.

The Advances in Knowledge Acquisition, Transfer, and Management (AKATM) Book Series (ISSN 2326-7607) is published by IGI Global, 701 E. Chocolate Avenue, Hershey, PA 17033-1240, USA, www.igi-global.com. This series is composed of titles available for purchase individually; each title is edited to be contextually exclusive from any other title within the series. For pricing and ordering information please visit http://www.igi-global.com/book-series/advances-knowledge-acquisition-transfer-management/37159. Postmaster: Send all address changes to above address.

Titles in this Series

For a list of additional titles in this series, please visit:
https://www.igi-global.com/book-series/advances-knowledge-acquisition-transfer-management/37159

Techniques for Coding Imagery and Multimedia Emerging Research and Opportunities
Shalin Hai-Jew (Kansas State University, USA)
Information Science Reference • ©2018 • 483pp • H/C (ISBN: 9781522526797) • US $235.00

Cross-Disciplinary Approaches to Action Research and Action Learning
Tome Awshar Mapotse (University of South Africa, South Africa)
Information Science Reference • ©2018 • 380pp • H/C (ISBN: 9781522526421) • US $215.00

Cross-Cultural Analysis of Image-Based Assessments Emerging Research and Opportunities
Lisa Keller (University of Massachusetts Amherst, USA) Robert Keller (Measured Progress, USA) and Michael Nering (Measured Progress, USA)
Information Science Reference • ©2018 • 166pp • H/C (ISBN: 9781522526919) • US $130.00

Social Media for Knowledge Management Applications in Modern Organizations
Francesca Di Virgilio (University of Molise, Italy)
Business Science Reference • ©2018 • 384pp • H/C (ISBN: 9781522528975) • US $215.00

Philosophical Perceptions on Logic and Order
Jeremy Horne (International Institute of Informatics and Systemics, USA)
Information Science Reference • ©2018 • 402pp • H/C (ISBN: 9781522524434) • US $225.00

Handbook of Research on Tacit Knowledge Management for Organizational Success
Dhouha Jaziri-Bouagina (University of Sousse, Tunisia) and George Leal Jamil (Informações em Rede Consultoria e Treinamento Ltda, Brazil)
Information Science Reference • ©2017 • 542pp • H/C (ISBN: 9781522523949) • US $275.00

Bio-Inspired Computing for Information Retrieval Applications
D.P. Acharjya (School of Computing Science and Engineering, VIT University, India) and Anirban Mitra (Vignan Institute of Technology and Management, India)
Information Science Reference • ©2017 • 388pp • H/C (ISBN: 9781522523758) • US $205.00

For an enitre list of titles in this series, please visit:
https://www.igi-global.com/book-series/advances-knowledge-acquisition-transfer-management/37159

IGI Global
DISSEMINATOR OF KNOWLEDGE

701 East Chocolate Avenue, Hershey, PA 17033, USA
Tel: 717-533-8845 x100 • Fax: 717-533-8661
E-Mail: cust@igi-global.com • www.igi-global.com

Editorial Advisory Board

Table of Contents

Detailed Table of Contents

The main objectives of this chapter are to define liquid workforce, describe its characteristics, and outline its implications. This chapter would help readers understand the need for studying liquid workforce and appreciate its importance in the contemporary world. Moreover, it provides a detailed outline of the ways and means that organizations can use to build such a workforce. It highlights the steps an organization can take to enhance the effectiveness of its liquid workforce. In addition, it explores certain best practices in the industry to illustrate how well the fluidity of the workforce can be managed without compromising on organizational goals.

Financing is an essential element for running any organization without which an organization cannot even be created. With many new organizations coming up and foreign companies entering into the market competition has touched its peak. This is valid even for scenarios where companies have to face a cut-throat competition to get the investors for themselves. New and booming startups have to compete with

as game-based learning, mobile learning, social learning, and virtual world learning, the teaching and learning environments are being extended. The evolution of high levels of sophisticated information technologies across the globe has tremendously improved intellectual capital formation through digital collaboration, and interactions. Therefore, it takes continuous update of intellectual assets through digitized processes to keep abreast of vast innovations and technical know-how.

Section 2
Impacts of Digitalization on Work Structures

This study has the aim of identify the main causes of a bad work environment with a high rate of turnover. The objective is to propose an intervention plan to increase the participation, commitment and employees' proactivity. This job is performed with a case study with the quantitative paradigm, transversal and exploratory; the selected sample is from a PYME dedicated to automation power services. For it is based on the model of situational leadership Hersey and Blanchard, in addition to job satisfaction survey NTP 213.

Digital work culture followed by any organization on the digital workplace in digital environment, work is going through the technologies in the virtual office, business site merely presents in cyberspace by using their laptop, cell phones via accessing the internet, anywhere, anytime. It directly affects economic, social, culture and business over the globe in a nano-second equally. It plays an important role in the development and growth of any organizations as well as the economic growth of the nation work culture is actually beliefs, ideas, manner and attitude of employees along with organization's ideologies and principles. author presents information about Digital media has impacting on Work Culture now traditional work culture is become digital work culture where efficiency with qualitative and quantitative productivity of employees and organization and business in a very swiftly, time saving, economically by the sharing information globally.

has been a rapid speed of new technological developments in the area of education. These new developments have created learning opportunities for working people. Digital technologies can now assist workers in their training and education needs thus making the worker of today more viable in the labor market. In the workplace, E-learning has made considerable inroads in the lives of workers development. E-learning is a combination of two important constructs namely learning and technology. Learning is a cognitive process where a student filters knowledge that he retains. Technology if used properly in this era can be used as a teaching tool to enable learning.

Social Networking is a good source to stay connected with others so it can be used as a tool in teaching. Those students, who are not able to learn the concept in the regular classroom, go to tuition. They work whole night to complete their homework and assignments meanwhile they cannot resist themselves from using social networking sites. They have a view that they can multitask, but actually, most of the concentration is in social networking and messengers and less in studies. So ultimately their studies suffer. Because of this, they feel unrest in next day classroom also. To change the habit of using the mobile phone and social networking is very difficult. Now they are addicted to using these. So, the only solution is to change the way they use these things. The only requirement is training the student about the use of these systems, for which they are made of. So, the basic aim of this research is how social networking can be used as an Efficacious tool in teaching and utilizing the time wastage on social networking sites and messenger by students in studies.

The present study is aimed at achieving main objectives i.e. to study the effect of Inquisitiveness and Attitude on Teaching Effectiveness among pupils' teachers practicing classroom-teaching practices; to study the effect of Inquisitiveness and Attitude on Teaching Effectiveness among pupils teachers practicing online-teaching practices, to explore the effectiveness of online teaching practice, to find out the future horizons of online teaching practice in Indian perspective. The sample of the present study is pupil teacher studying in Dayalbagh Educational Institute Deemed

Foreword

Companies are attempting to adopt the latest technologies to address the issue of competitive differentiation. The evolving technology investment environment has a deep influence and impact on how organizational structures, value chain collaborations, internal operations, and customer experiences for both products and services, will emerge in the immediate future.

There are several ways to create differentiators in a competitive and a dynamic market, but above all, the establishment of the 'digital enterprise' holds immense promise in bringing about the next wave of product innovation and business process transformation. The combination of next-generation technologies like cloud computing, big data, advanced analytics, social media marketing, mobility, voice of customer management and e-commerce are slated to change the very basis of engagement across businesses. As top business leaders are increasingly turn to digital technologies, organizations that make the most of this opportunity will be at the forefront of the industry. Thus, in the near future, 'digital' and 'business' will be nearly synonymous for organizations.

This comprehensive text aims to get insights into how the technologies are influencing the very structure of workplaces and various workplace practices. This book successfully provides theoretical frameworks, the latest empirical research, problems and future research prospects in the dynamic and fast-evolving digital ecosystem. The present text is divided into four sections and contains 11 chapters. The issues covered are the liquid work force, crowd funding, e-learning and the effects of inquisitiveness and attitude on teaching effectiveness among students and teachers, the re-shaping human capital formation through digitization, the impact of digitization on work culture, and the emerging trends of social networking and its effects.

A noteworthy and invaluable feature of this book is the large number of chapters drawn from a variety of situational and necessary components to help the readers to understand the concepts and applications of different aspects of digitization. The current wave of digitization is a result of the power and impact of converging technologies like big data, enabling enterprises to explore new business models and differentiation opportunities, both on the production side as well as the customer engagement side of the business. The challenges of digitization in organizations are both difficult and interesting. People are working on them with enthusiasm, tenacity, and dedication to develop new methods of analysis and provide new solutions to keep up with the ever-changing threats. This book is truly a valuable addition in this direction.

The editors, Punita Duhan, Dr. Komal Singh, and Rahul Verma, have made a sincere attempt to cover various facets of digitization as envisioned by practitioners. Valuable insights can be found by reading the whole book.

I compliment all the editors for this academic venture, so that academicians, researchers and policy makers may benefit from this presentation.

S. K. Singh
L. N. Mithila University, India

Preface

We are witnessing a fresh acceleration of technological change which is affecting the established ground rules within companies and across industries alike. In this unpredictable scenario, Big Data is set to play a key role. Digital technologies are in fact assisting workers and making them even more viable in the labor market. For businesses, the next wave of work redesign has great potential to improve both the efficiency and effectiveness of the most expensive talent in their organizations. For policy makers, the continuing transformation of jobs by technology means that worker's skills need to evolve even more rapidly. In other words, software and other things or devices that assist and guide workers, help them become creative and innovative. A relatively recent and comprehensive study on the future of employment in 702 occupations concludes that, "For workers to win the race, however, they will have to acquire creative and social skills." One of the most important skills that all workers need to cultivate is innovation.

Adaptive Digital Enterprises need to manifest a platform that continuously learns, assists, and guides its workers to be more creative and innovative. The very first trend – namely innovation – remains the most important. Leveraging a "Build for Change" platform with the associated creativity disciplines, an adaptive digital enterprise can focus on innovation especially within their processes, customer experiences, ideation for new products, and a modernized IT organization.

Digitalization will remake employee-enterprise connectedness in the same way that it is revolutionizing customer-enterprise relationships. Going digital enables more collaborative and comprehensive employee engagement, which in turn drives higher levels of employee performance. Faced with massive change, some leaders will struggle simply to stem the tide without fundamentally changing their organizations. Others will look for ways to jump ahead of the curve. Leaders will need a framework for digital enterprise transformation that drives a holistic digital strategy in the four quadrants of transformation *viz.* Product and service innovation, Value ecosystem, Go-to-market and Culture and organization.

THE CHALLENGES

Organizations face enormous challenges in the digital age since old rules no longer apply. As content and consumption become increasingly digital, old order business models based on tight control of distribution of premium content—to the point of scarcity—are no longer relevant. Consumers are increasingly demanding products and services in multiple ways, across multiple platforms, and at a time that is convenient for them.

In the digital arena, every element of the business process engineering is being questioned or redefined. The whole processing cycle can be cut down to a few weeks as against months in the traditional sources, thereby reducing cost and increasing revenue. The old order is being replaced by a new one where different attributes are valued. As we shift from a distribution paradigm to a consumption paradigm, companies have to rethink their strategies. The balance of power is shifting from the content owners to platforms and from producers to consumers. We are moving from a B2B to a B2C environment and potentially a Creator to Consumer (C2C) environment. The implications of the shift from a distribution "push" to a consumption "pull" paradigm in an entirely by-table world are truly revolutionary.

Key changes the industry can expect in the future:

- Technical fluency will be essential in all aspects of the new digital enterprise.
- Delivery of content in multiple forms and contexts will be a far more critical component of success than it is today. Hence, the ability to manage increasingly complex product "fragments" and "bundles" will be crucial for business systems.
- The ability to manage the associated rights and royalties will become much more complex.

SEARCHING FOR A SOLUTION

We are in the era of high-tech, digitalized and modern civilizations. In the current worldwide economy, the use of digital technologies is served as a means of boosting economic activities. The mass adoption of digital technologies through connected services and devices has proven to accelerate economic growth and facilitate job creation. Build training as a core competency in the organization to actively generate skills that are in demand and unveil new talent in the organization. Every job is digital job; technology is fundamentally changing the way people work and the work they do, a trend in businesses embraces as they move to compete in the digital age. Train your own; companies must take an active role in creating the skills they need by making training a critical component of their talent management strategy.

ORGANIZATION OF THE BOOK

The book is organized into four sections, namely "Reorganization of Work Structures Through Digitalization" consisting of four chapters; "Impacts of Digitalization on Work Structures" consisting of two chapters; "Digitalization: Trends and Concerns" consisting of two chapters; and "Digitalization: Metamorphosis of Educating and Mentoring" consisting of three chapters. A brief description of each of the chapters is as follows:

Chapter 1 establishes to define liquid workforce, describe its characteristics and outline its implications. This chapter would help readers understand the need for studying liquid workforce and appreciate its importance in the contemporary world. Moreover, the author provides a detailed outline of the ways and means that organizations can use to build such a workforce. The author also highlights the steps an organization can take to enhance the effectiveness of its liquid workforce.

Chapter 2 establishes the need for new methods of financing such as crowd funding as an essential element for running any organization without which an organization cannot even be created. The chapter sets the scene for discussions as to how Crowdfunding has been positioned as a viable option in line with traditional methods of financing such as venture capitalism and angel investments. In particular, the chapter identifies that Crowdfunding has gain popularity amongst new and inexperienced entrepreneurs and for getting money to carry out social, cultural and environmental projects.

Chapter 3 presents a review on how is crowd shaping affecting the life of a retail shopper in this technical savvy world. The author identifies the existing challenges and difficulties faced by the business in today's era where change is the only constant. The author contends that the innovative and improved technologies transform the industries and companies and change lives. It also identifies that the companies continue to invent new products and technologies to cater the future needs of customers.

Chapter 4 focuses on reshaping intellectual capital formation via electronic-based learning platforms. The author critically examines the literature on human capital development through e-learning and finds that digitalization enhances teaching/learning processes and activities, rather than rendering the traditional methods obsolete. In particular, the chapter identifies the evolution of high levels of sophisticated information technologies across the globe that has tremendously improved intellectual capital formation through digital collaborations and interactions.

Chapter 5 aims to identify the main causes of a bad work environment with a high rate of turnover. The objective of the chapter is to propose an intervention plan to increase the participation, commitment and employee pro-activity. The author accomplishes his task by performing a case study with the quantitative paradigm, transversal and exploratory.

Chapter 6 assesses the impact of digitalization on work culture with an outcome of qualitative and quantitative productivity of employees and business by sharing information globally. The author grounds her argument that the digital workplace in digital environment plays an important role in the development and growth of any organization as well as the economic growth of the national work culture.

Chapter 7 analyses and compares recent trends for digitalization in the organizations. The author systematically reviews the cut-throat competition in the today's world and presents the need of consistently working towards the invention of such gadgets which will reduce their time so that they can manage their activities effectively. The author also identifies the importance of Digitalization in shaping the workforce and work environment of any organization.

Chapter 8 analyzes how various strategies followed by Uber have helped it to become a globally successful player. The author examines to see Uber as a technology-based company. The overall aim of the chapter is to consider the different strategies that take Uber to join the global market successfully.

Chapter 9 presents an analysis of the impact of digital technologies on the world of work and its environment as technology continues to permeate all areas of society. The author grounds her arguments that there has been a rapid speed of new technological developments in the area of education and these new developments have created learning opportunities for working people. The author contends that E-learning is a combination of two important constructs namely learning and technology and assesses the impact E-learning has made on the lives of workers development.

Chapter 10 analyzes on how social media can be used as an Efficacious tool in teaching and utilizing the time wasted on social networking sites by students. The author presents an array of tools and techniques that help in training the student about the use of these systems, for which they are made of. The authors argue that Social Networking is a good source to stay connected with others so that it can be used as a tool in teaching and so the only solution is to change the way they use it.

Chapter 11 is aimed to study the effect of inquisitiveness and attitude on teaching effectiveness among pupil-teacher practicing classroom teaching. Also, the author explores the effectiveness of online teaching practice and finds out the future horizons of online teaching practice in Indian perspective. The chapter sets the scene for discussion that pupil-teachers possess the more positive attitude and inquisitiveness towards online teaching practice in comparison to classroom teaching practice. In particular, the chapter identifies that inquisitiveness and attitude significantly predict the teaching effectiveness in online teaching practice and classroom teaching.

In nutshell, through this volume, editors have tried to capture the unprecedented and unparalleled experiences encountered by policy makers, leaders, entrepreneurs and organizations while trying to incorporate the emerging digital technologies in their work spheres. Book is a sincere attempt to encapsulate the novel ways and means adopted by researchers throughout the world to cope up with this digital disruption. Editors believe it will be a good value addition to the existing body of knowledge.

Punita Duhan
Meera Bai Institute of Technology, India

Komal Singh
Meera Bai Institute of Technology, India

Rahul Verma
Bhai Parmanand Institute of Business Studies, India

Acknowledgment

A book is never the work of only one or two people. Getting this book into your hands has also been a team effort. It took the painstaking efforts of an editorial board, reviewers, research scholars, editors and the publisher to put this volume together.

Editors owe a great deal to the masters of the subject and researchers whose work has been consulted and drawn upon in the course of discussion of this book. Editors express their gratitude to the contributors without whose contribution, it was impossible to bring out this volume. Editors will fail in their duty if the comments, compliments and suggestions of reviewers and editorial board are duly acknowledged. Further, heartfelt gratitude also goes to the publisher, IGI Global, without which it was not possible to help this volume reach the readers.

Last but not the least, the editors want to thank their families for their unconditional support to help them complete this book in time.

In the end, we express our gratitude to the reader for the attention given to the book.

Punita Duhan
Meera Bai Institute of Technology, India

Komal Singh
Meera Bai Institute of Technology, India

Rahul Verma
Bhai Parmanand Institute of Business Studies, India

Section 1

Reorganization of Work Structures Through Digitalization

Chapter 1
Liquid Workforce:
The Workforce of the Future

Manish Gupta
IBS, IFHE University, India

ABSTRACT

The main objectives of this chapter are to define liquid workforce, describe its characteristics, and outline its implications. This chapter would help readers understand the need for studying liquid workforce and appreciate its importance in the contemporary world. Moreover, it provides a detailed outline of the ways and means that organizations can use to build such a workforce. It highlights the steps an organization can take to enhance the effectiveness of its liquid workforce. In addition, it explores certain best practices in the industry to illustrate how well the fluidity of the workforce can be managed without compromising on organizational goals.

INTRODUCTION

Creating an efficient and effective workforce has been the top priority of the HR function for decades (Sidibé & Campbell, 2015). However, to catch-up with the ever-changing business needs and dynamic corporate environment, the ways of managing workforce have undergone a sea-change since the pre-classical era (Hoang, 2015). Organizations in the modern era are digital, extremely competitive, global, and highly result oriented. In such a scenario, managing the workforce effectively is the best way to retain competitive edge and remain globally relevant (La Rosa, 2016). Whereas the digitization of most of the HR activities has made HR function simpler and efficient, it has also necessitated a turn-around of the current HR practices

DOI: 10.4018/978-1-5225-3191-3.ch001

to address the dynamism in today's workforce (McFarland & Jestaz, 2016). This chapter mainly describes the concept of liquid workforce and how organizations can embrace the dynamic work environment by using their workforce.

Need for New HR Practices

A survey report 'Technology Vision 2016' by Accenture using a sample of more than 3,100 business and IT executives revealed that 33% of the participants feel that the world economy is already affected by digital revolution (Accenture, 2016). It also reported that 86% of the respondents predict a sharp rise in the pace of technological changes over the next three years and organizations are increasingly investing in the tools that help organizations adapt to the dynamism of the digital era. Automation is increasingly taking over manual tasks and a whole lot of millennial are changing the current workforce with their digital insights (Zaino, 2005). Some startling facts also reaffirm the urgent need for embracing these changes. For example, around 43% of the current US workforce is likely to be freelance by the end of 2020 (Neuner, 2013). Practitioners across the globe find its remedy in developing liquid workforce which has certain inherent ability to adjust itself to the highly dynamic business environment. Four major reasons behind the rise of liquid workforce including a steep rise in the number of freelancers, increasing automation of mundane activities, improving the rate of innovation, and entry of 'millennial' into the workforce (FastTrack Press Room, 2016). These reasons are summarized below:

- Increasing automation of mundane activities: There are certain activities in any job that are too repetitive and labor intensive that can be carried out by the machines. Reducing human intervention in such activities is not only cost effective for the organization but also beneficial to the employees as it helps employees utilize their talent in more productive activities. A recent example is of India's ICICI bank introducing software robotics in 2016 with a target of automating 20% of its transactions by the end of the year 2016. The level of digital war can be smelled from the very fact that this decision by the ICICI bank was preceded by an announcement by HDFC bank that it would use physical robots at its branches (Nair & Gaur, 2016).
- A steep rise in the number of freelancers: As the economies are embracing digitization, there is a growing demand of freelancers who are easily available to work on contract. Hiring a freelancer is a win-win decision for the organization and the freelancer. It is because while freelancer can focus his/her attention only on the particular job mentioned in the contract, the organization would only have to pay for the limited time the freelancer worked and needs not to bear the amount that a full-time employee would

otherwise take. Several websites such as 'https://www.freelancer.in' bridge the demand-supply gap between the organizations and the freelancers.

- Improving the rate of innovation: It is no longer sufficient for employers to limit themselves to innovations. Instead, they need to go a step ahead and enhance the pace of innovation to survive and thrive in the intense competition. Recently, cognizant carried out a survey on the future of talent and digital skills using a heterogeneous sample of 422 European and U.S. executives (Cognizant, 2016). The findings revealed that majority of executives believe that there is a digital skills gap from "moderate" or "severe." This gap restrains their organizations from realizing their digital dreams. The report also suggests that innovation can be fostered by linking proprietary platforms to 'talent clusters'. It is the transient workforce that can quickly adapt and add value to business reinvention. The company management needs to come forward to junk the old big-sized talent clusters to lean and small clusters that serve a niche.
- Entry of 'millennial' into the workforce: The term 'millennial' or more specifically 'millennial workforce' generally refers to a set of people which entered the workforce 2005 onward. According to a PWC survey 'Millennial at work: Reshaping the workplace' in the year 2011 using a sample of 75 countries, the changing the workplace is all set to change because millennial generation matters (PWC, 2011). It is because the millennial generation is more tech savvy, self-oriented rather than job-oriented, socially connected, anxious for rapid progression, flexibility seeking, and willing to scale new highs. Google and Apple are the pioneers in attracting and retaining most talented millennial because these companies believe in encouraging the employees to think out of the box. To retain such a workforce, processes need to transparent, targets need to appropriate, the doors of opportunities in participating strategic decision making must be opened, strategies should be millennial specific, and regular feedback about the current position and how it can be improved must be preferred over negative feedback (Gilbert, 2011).

It is evident that the way employees work is all set to change yet again. Gig economy, a kind of work environment in which most of the jobs are on short-term contracts, is swiftly taking over the present situation (Goodwill, 2016). Thus, it is of paramount importance to know how the shift from the present 'static' to the next generation 'dynamic' workforce would take place (Simons, 2016). Such shifts are captured in five points including business function to formal online crowd funding platform alignment, ad-hoc to continuous training, segregated to end-to-end workforce management, concentrated to participative innovation, and low to high collaboration as described below:

- From business function to formal online crowd funding platform alignment: Crowd funding is an online collaborative funding from peers. The importance of this platform has also been highlighted by the Forbes magazine, "Unless you've been living in a remote island for the last few years, you've heard about crowd funding or stories of people raising thousands or millions of dollars online" (Barnett, 2013). Some of the crowd funding websites includes Kickstarter, Indiegogo, Rocket Hub, FundRazr, and GoGetFunding (Taylor, 2013).
- From ad-hoc to continuous training: 'Ad-hoc' is a Latin word which means 'for this'. Instead of ad-hoc training which focuses on shaping an employee's knowledge, skills, and abilities for a particular project or job, the dynamism of the demands requires training on enhancing skills that can enable employees to adapt to the changing environment. Giving continuous training is beneficial to the company in many ways. First, it helps employees manage their daily tasks. Second, it makes employee productivity almost immune to the changes in the external environment. Third, it enhances the self-confidence of the employees. Fourth, it helps employees achieve their goals. Continuing Professional Development (CPD) offered by Chartered Institute of Personnel and Development (CIPD) is a flexible process that trains and provides employee feedback online continuously.
- From segregated to end-to-end workforce management: While traditional management tools rely on fragmented workforce management such as tracing the number of hours an employee works and adjusting the pay accordingly, evaluating the availability of staff members and their qualifications for several related HR activities including preparing or revising schedules and distributing these schedules, handling leave applications, and preparing annual HR reports. Such fragmented activities mostly prove to be erroneous, costly, and time consuming. Consequently, it is hard to work on one specific area or issue without any assistance from a tool or management solution. Nevertheless, installing a comprehensive workforce management solution that helps the HR managers to track a particular strand or a sequence of related activities seamlessly from beginning to end without compromising on the efficiency (Amodio, 2013).
- From concentrated to participative innovation: Generation Y (the millennial generation) believes in contributing fresh ideas that can in some way add to the digital revolution. It would be extremely discouraging for this generation if innovative ideas are invited only from a particular department. This generation asks: 'Why not us?' The youngsters want to take part in shaping the company's future. It is something that boosts their self-esteem, makes them happy and gets them the feel of being valuable (Emelo, 2011; Kramer,

2010). The underlying concept is to allow any employee to submit, review, and rank an idea. Companies can encourage participative innovation, for instance, by facilitating a physical / online suggestion box for new ideas (Phillips, 2010). It is equally important for the concerned manager to go through those ideas and appreciate or reward the best one. Not only corporate sector but also governments are increasingly recognizing the potential of participative innovation. For example, a recent initiative of the Government of India to invite innovative ideas through public participation online is: 'https://mygov.in'.

- From low to high collaboration: Businessmen agree that collaboration is the need of the hour as it increases the productivity due to synergetic effects. One way to increase collaboration is by introducing electronic meeting systems (EMS) in the company (Rittgen, 2010). These systems decrease the requirement of attention and increase performance. The two advantages of e-meeting systems over face-to-face meeting are: (1) simultaneity which allows multiple people share their ideas simultaneously and (2) anonymity which means the utterances are not attributed to a particular speaker. Both these qualities coupled with reduction in the hassle of arranging a conventional meeting gives more room to the members to share their ideas which would not have been shared otherwise. With an improvement of 55 percent in group performance, the implementation of EMS in IBM has helped it make better use of its teams (Grohowski, McGoff, Vogel, Martz, & Nunamaker, 1990). IBM found that anonymity was beneficial, number of participants in the meeting increased, the engagement level of participants also increased, and fewer meetings over less time were required, and focus of the participants on the task increased. However, it needs special attention in some areas such as increased importance to pre-meeting planning, distribution of the minutes of the meeting and proceedings after the meeting, requirement of tech-savvy users, flexibility of the software to meet the group application variety, etc.

WHAT IS LIQUID WORKFORCE?

IBM coined the term 'liquid workforce' where "people are hired on an "as-needs" and "just-in-time" basis, for specific projects activities, whether for days, weeks, or months. Beyond, the conventional contractor model, the liquid workforce is underpinned by crowd sourcing, drawing together skills of freelance workers and the task need of organizations" (Bakhtiyari, Barros, & Stewart, 2014, p. 18).

Industry needs a workforce that can be 'seasonally' attached and detached without with ease as per the contingent strategic and operational needs. To deal with contingent requirements, contingent workers can be a critical tool to face the stiff competition and optimize business profits. The underlying idea behind the concept of liquid workforce originates from the belief that technology not only disrupts but enables business transformation. It converts a rigid and docile firm into a change-ready and agile firm. To illustrate, General Electric has recently introduced 'Fast Works', a technique for entrepreneurs for developing "The Lean Startup" (Power, 2014). It focuses on products that ensure fast learning and quick deliverables. It provides more autonomy to the users to swiftly alter direction on projects. Apparently, liquid workforce which is also termed as 'fluid workforce' is the need of today and the future. To gain further insights, the next section narrates some of its differentiating features.

THE KEY CHARACTERISTICS OF LIQUID WORKFORCE

Like any workforce, liquid workforce does have a set of defining features. Some of these characteristics are diversity in terms of composition, mobility in terms of workplace, flexibility in terms of working hours, accommodative in terms of inviting different skill-sets, open mindset in terms of embracing change, and ability to quickly learn, multitask, and adjust itself to meet the organizational objectives (Ethos BeathChapman, 2016; Pegasystems, 2016). These are detailed below:

- Diverse in terms of composition: The workforce has variety. Here, variety means different types of workers are included. In other words, the liquid workforce is inclusive as it includes employees other than the full-time staff. These could be part-timers, causal workers, and freelancers. A combination of these diverse types of workers helps companies accommodate dynamism in project demands. At the same time, such heterogeneity increases its efficiency.
- Mobile in terms of workplace: Another defining feature of such a workforce is the flexibility with which work can be handled from anywhere. The work can be completed beyond the boundaries of the office. Mobility of work helps employees strike a balance between their work and their personal life. Recently, ICICI bank has launched the program wherein work from home is allowed. It is beneficial to the company as it reduces a lot of company's office expenses.
- Flexible in terms of working hours: Flexibility in work timing is what millennial want. They would like to enjoy in the free time and work whenever they are like too. Forcing them to work from 9 to 5 does not work.

- Accommodating nature in terms of inviting different skill-sets: The liquid workforce is not rigid and is willing to work with people of diverse skill-sets. It is also an opportunity for the members to adopt their skills as they also need to exhibit multi-task behavior.
- Mindset in terms of embracing change: Members of liquid workforce are highly curious. It is in their blood to continuously learn from the dynamic environment and keep themselves open to extensive learning.
- Ability to quickly learn, multitask, and adjust itself to meet the organizational objectives: All the above qualities makes liquid workforce quick to change and re-direct its efforts towards renewed goals flawlessly.

DEVELOPING THE LIQUID WORKFORCE

Only because liquid workforce relies more on contingent employees does not imply that the management of the company can choose to be rigid and less-responsive to the changing environment. Transformation of conventional workforce into the liquid workforce is a strategic decision and full support from the management is needed to facilitate the process.

Though there could be different ways to develop a fluid workforce, Accenture prescribes a global procedure that can be followed by any type of organization. The specifics of this procedure can be decided by the organization based on the resources available at their disposal. Accenture offers this procedure in the package of 100 and 365-day plans (Accenture, 2016; McGuigan, 2016).

The focus of the 100-day five-step plan is on highlighting the need for change as well as on providing evidence that it works. Each of these five steps is described below -

Step 1: Conducting A Skill - Gap Analysis: The open roles need to be reviewed by the HR. Also, it is important to identify the roles on high priority that are still seeking suitable talent.

Step 2: Formulating A Renewed Training Strategy: A proper strategy needs to be put in place to ensure that organization's training capabilities can be enhanced. For this, the management of the company must decide on the investment required for providing the employees with consistent and effective training across the organizational hierarchy.

Step 3: Developing A Renewed Engagement Strategy: Because millennial have unique requirements, the organization needs to develop a proper engagement plan. The plan must give room to these millennials to work on what they are most passionate about and provide sufficient autonomy to boost their morale. If their roles complement their knowledge, skills, and abilities, the productivity of the employee is likely to improve and the talent can be engaged more efficiently.

Step 4: Preparing Strategy For Engaging The Freelancers: Apart from the full-time employees, the part-time employees (mostly freelancers) also need to be engaged. Therefore, a strategy which distributes the work clearly between these two types of workforce is of paramount importance. The strategy should ensure that there is proper communication between the short and long-term employees wherever necessary.

Step 5: Piloting A Fresh Liquid Project: Transforming the workforce from conventional to liquid is expensive. Therefore, it is advisable for the organizations to predict the possible impact of such kind of transformation on the productivity and explore the possible contingencies to reduce the amount of operational risk involved. In order to conduct a pilot test, a company needs to choose one of the liquid groups and assign it a project that has autonomy and necessary resources for meeting the objective of the project. The results of this step would provide the necessary foundation for the complete workforce transformation.

Unlike the 100-day plan, the focus of 365-day plan is on implanting the pilot liquid project across the organization. The 365-day plan is followed once the 100-day plan is successful and it also espouses five steps. These steps are discussed in the following paragraphs:

Step 1: Putting The New Training Strategy To Work: Analyze the result of the pilot test conducted in the 100-day plan to select the optimum combination of training the different sources such as personalized training, MOOCs, or boot camps. The combination depends on the context a company operates in.

Step 2: Using The Experience Of The Pilot Group: By now, the company has tasted the results of the pilot test and has discovered an optimal combination suitable to its context. The next step is to formulate a plan specific to extending the practices followed during the pilot liquid project exercise across the organization. For this, the company needs to meticulously look into the strengths and weaknesses of the pilot projects, learn from the mistakes, and build a robust plan that can face the dynamic external environment to a greater extent.

Step 3: Planning To Scale Your Liquid Project Practices Across The Whole Company: It is important for the companies to know those areas, ways, and means that can benefit the most from crowd sourcing. In order to accomplish this objective, the company needs to start three projects simultaneously. Each of which has a focus on a different area that can be addressed well by the crowd. The companies should rely on the results of these projects for sanctioning the crowd sourcing projects for the rest of the company or those parts of the company that are likely to be benefitted the most.

Step 4: Integrating HR Analytics With The Organization: The next step is to integrate all the HR activities such as, recruitment, selection, promotion, compensation, retirements etc of the company with the liquid workforce. It can be done with the help of data scientists who can apply HR analytics to gain new insights from these conventional activities. All the gaps must be bridged to attract and retain talent.

Step 5: Formally Introducing Predictive Analytics To One Area Of People Management Strategy: The company should strive hard to develop a proper plan to understand how predictive analytics could enhance existing HR practices. This plan needs to be comprehensive and foolproof to the extent possible.

Aligning Liquid Workforce With Business Strategy

Organizations often find it difficult to face challenges arising out of the mismatch between business strategy and a giant operational shift. It is essentially because of not following a fool-proof alignment procedure that can ensure that the strategic shift turns out to be a success. Specifically, the organization must change on certain parameters before turning its workforce to a liquid workforce. These are as follows (Wells, 2016):

- **Introducing Context Specific Training Across the Organization:** As per the report published by Accenture, the next generation work would be flexible or fluid workforces and to align such a workforce with the business strategy would require business projects with embedded training. This training should not only be relevant to meeting the objective of the project but also of some relevance to the individual learner. For instance, if the company hires an external expert say a freelancer, the company would want to prepare a training program that focuses not only on unique organizational processes but also on industry material. In this way, alignment of the organizational specific processes can be done with the industry. Another important way could be seamless distribution-and redistribution of data and information among the concerned teams or members. For this, the company must grant leverage to how the exchange of content is conventionally done. The company needs to explore cloud-based options to store, retrieve, and exchange data from anywhere by any of the concerned team member. It would reduce a lot of redundancy. Such kind of training is provided by third parties such as Mind flash.

- **Ensuring Collaboration in the Distributed Workforce:** It is indeed difficult to ensure that all the team members cooperate, coordinate, and collaborate to accomplish a common goal. It becomes more challenging when the collaboration has to be done in real-time. The possible way out is to facilitate right communication among the members using right tools. There needs to be dedicated slack channels and spaces for that. Though it would require initial capital investment, a lot of money relating to commuting can be saved once such system is developed and is fully functional. The company can also opt to outsource this activity of keeping all connected to each other and providing the team members a common platform to collaborate.
- **Measuring Team and Project Effectiveness in Frequent but Short Meetings:** Finally, it is equally significant to appreciate the importance of feedback and develop mechanisms to capture and analyze it quickly. It directs the team on time so that efforts and resources are not wasted. It can be done by following 'Agile Scrum framework'. Scrum is a framework that details how a product can be managed. It relies on the philosophy that customers' demands are volatile. This initiative coupled with a 15-minute team meeting on daily basis in which even distantly related or peripheral project objectives can be discussed and the pre-set milestones can be refined further as per the need of the situation. If handled with utmost care and given support from the top management can prove to be a game-changer for the company. Such meetings are called as micro meetings as the time period for each meeting is too small to get bored. A company can also choose to integrate its cloud-based content and learning management system with its customer relationship management (CRM) system. There are companies that implement such high-tech solutions. For example, TruAssets, a national property management company has integrated its course work relating to contractors and results of the quizzes with its CRM portal. Consequently, the company can now trace, handle, and assess its training related results for its several external vendors. General Electric is another big company that has formed Fast Works teams to make the use of liquid workforce. The details of the same are provided in the next section.

LEARNING FROM THE BEST PRACTICES

Though the concept is interesting and has promising outcomes, only a few organizations took the risk of implementing it and others chose to wait-and-watch. One of the few organizations that have set benchmarks of building a liquid workforce successfully is General Electric as detailed in the subsequent paragraphs:

The Case Of General Electric (https://hbr.org/2014/04/how-ge-applies-lean-startup-practices)

Brad Power, a well-known consultant, narrates the story of General Electric. According to him, the company, General Electric (GE), is one of the first companies that begin tapping the potential of agile and fluid workforce. In case of this company, the initiative to align with the dynamic requirements of the customers was taken in the form of 'Fast Works'. It is a framework that helps entrepreneurs develops 'The Lean Startup', an approach for developing fresh products that come out of supple software development and ensure swift deliverables and quick learning. Interestingly, it is not for an information technology (IT) activity. Since GE is a manufacturing company, it wanted to match its operational activities with the quick learning cycles of the customers. In a way, the company wanted to milk its appliances fully before these get obsolete in the minds of the customers and enter in the dog phase for the company. It was a bold decision by the top management to re-formulate the operations strategy completely to suit the requirement of the customers.

GE is a pioneer in taking giant and mostly risky steps to adapt itself to the external world. For example, in the year 2008, GE invested USD 1 billion to transform its existing plants, buy new ones, and recruiting 3,000 new employees. In the year 2013, the Chief Executive Officer (CEO) of the company, Chip Blankenship, challenged his newly formed cross functional team by saying: "You're going to change every part the customer sees. You won't have a lot of money. There will be a very small team. There will be a working product in 3 months. And you will have a production product in 11 or 12 months." What the team was allowed to do is interesting and inspiring.

The team members were given a separate room so that the boding among them can be improved. As a result, it became a tight group. The members used to go down to the factory floor an build the products collectively. Unlike the conventional approach in which the customer or market requirements are given by the sales force, this team was in direct and constant touch with GE customers. The customers were also the participants in the development of the products throughout. Hearing the feedback about the product directly from the horse's mouth (the customers) was a refreshing and exciting experience for the engineers who otherwise work in the back office. However, receiving customer feedback directly was also a challenging task. The team had to bounce ideas and product prototypes off of retail salespeople who wanted to know about GE's products in Louisville, GE's training centre. It also had to visit several product design centers located in Chicago as well as in New York.

The objective of visiting these centers was to learn the details of the products for the respective clients. The feedback wasn't a sweetener to the team members' ears but it laid a concrete foundation of change. The team soon came out with a new idea 'minimum viable product'. The product was a refrigerator with fresh styled doors that is, it opens from the middle. Started in January 2013, the product went through several rounds of corrections based on customer feedback relating to the darkness of the steel, lighting, etc. It was only by August, 2013 that customers actually started liking the refrigerator which was version 6. The team readily started working on the next version i.e. version 7 and decided to come out with improved version every year. It was a radical shift in the strategy of GE as the company earned its fame by introducing a revised product in the market every five year. Till the launch of the revised product, the new product features used to be kept secret. Kevin Nolan was quick to say, "With Fast Works we're learning that speed is our competitive advantage. How do we become much more open and collaborative with the customer base? You can't do that if you want to be secretive."

The company made several changes so that Fast Works can actually function. The company re-worked majorly on three areas namely supplier relations, finance, and roles and responsibilities. Each of these is described below in short:

Supplier Relations: Since GE is a manufacturing company; the suppliers play an important role in delivering the required product to the customer on time. The team was well aware about this fact and decided to approach the suppliers and make them understand what exactly they are doing and how it would affect the supply chain. As expected, the suppliers were extremely grateful to the company that they were given a chance to participate and get involved in such a major exercise.

Finance: In the words of Vic Roos, Lead Purchasing Program Manager: *"We let a finance guy in the room. He helped us challenge the big company mentality. At times we moved much faster than the company would normally allow. At times it drove the materials manager crazy."* It shows the kind of autonomy granted to the finance person and the open-mindedness with which the suggestions of the finance person were heard. Certain findings were shocking such as the company needed to have annual or bi-annual payback but that kind of model cannot survive in the situation when the things are extremely dynamic and the future demands of the customers are uncertain. Obviously, it is extremely difficult to evaluate the cost of such a project. When the company first evaluated the ongoing project, the financial numbers looked really bad. But the team manager decided not to worry about what the investments or expenses incurred on the entire program for calculating payback. This take of the manager relieved the team members from financial pressure and they could focus on the execution of the project rather than worrying about the cost it may incur. The opportunity costs relating to speed are not assessed by the traditional financial systems.

Such costs are incurred due to not tapping the potential customer who could have otherwise purchased the product if delivered on time.

Roles and Responsibilities of the Leader: Of course, there was a full support not only the financial support but also moral support from the top management. There were instances when there was a mismatch between the expected and actual speed of the project. At that time, the CEO conducted a meeting and resolved the issue. Also, a lot of autonomy was granted to the team in terms of freedom from taking several approvals for pivots. Full autonomy was given to the team to take such decision on its own.

The outcome of this entire exercise has been fruitful to the company. The findings indicate that it could cut the program cost to only a half and could double the program speed. Moreover, the product started selling at more than the double rate of the normal rates. The company is now working on extending these practices to other appliance manufacturing units as well. For instance, GE Appliances hosted 7080 young GE high potentials. Furthermore, 80 coaches were trained and were dedicated exclusively to Fast Works. Around a thousand GE executives were exposed to Lean Startup and another giant step was the launch of over 100 Fast Works projects worldwide.

DIGITIZATION FUELS LIQUID WORKFORCE

Thanks to the last global recession, the companies have realized that keeping only full-time permanent employees is not a sustainable model. It is because liquid workforce facilitates faster change compared to the illiquid one. Not surprisingly, the market of freelancers, part-time employees, and casual workers is all set to thrive. Increased level of digitization across the globe is boosting the prospects of freelancers. It is a way to create one's own job. There are several examples of such freelancing digital marketplaces. These include Fiverr, Etsy, UBER, Threadless, Kaggle, Firefox, Toptal, HourlyNerd, Tispr, Teach Me Now, Engineer it etc. (https://innovationlab.net/blog/freelance-economy-liquid-workforce)

CONCLUSION

Now that the world is increasingly getting digitized, most of the organizations are feeling the need of doing away with the traditional workforce management practices that are no longer relevant and effective. Some of the organizations such as Accenture,

Google, and GE have already set the examples of outperforming competition by building liquid workforce, a necessity to build inherent competency of adapting itself as per the dynamic needs of the stakeholders.

REFERENCES

Accenture. (2016). *Accenture Technology Vision 2016: People First: The Primacy of People in a Digital Age*. Retrieved from http://goo.gl/JGfQvJ

Amodio, M. (2013, April 08). The Challenges of Workforce Management and How to Triumph Over Them. *Workforcemanagementtoday*. Retrieved from http://www.workforcemanagementtoday.com/articles/333395-challenges-workforce-management-how-triumph-over-them.html

Bakhtiyari, A., Barros, A., & Stewart, G. (2014). *Applying Enterprise Architecture to Business Networks*. Retrieved from http://eprints.qut.edu.au/73012/1/73012.pdf

Barnett, C. (2013). Top 10 Crowd funding Sites for Fundraising. *Forbes*. Retrieved from http://www.forbes.com/sites/chancebarnett/2013/05/08/top-10-crowdfunding-sites-for-fundraising/#3e9d76eb1cfb

Cognizant. (2016). *People — Not Just Machines — Will Power Digital Innovation*. Retrieved from https://www.cognizant.com/whitepapers/People-Not-Just-Machines-Will-Power-Digital-Innovation-codex1850.pdf

Emelo, R. (2011). What if millennial ran your mentoring program. *Chief Learning Officer*, *10*(5), 32–36.

Ethos BeathChapman. (2016, July 19). *The liquid workforce: A well of innovation*. Retrieved from http://ethosbc.com/industryinsights/theliquidworkforceawellofinnovation

FastTrack Press Room. (2016, July 14). *How can you facilitate the liquid workforce?* Retrieved from https://www.fasttrack.com.au/blog/howcanyoufacilitatetheliquidworkforce

Gilbert, J. (2011). The Millennial: A new generation of employees, a new set of engagement policies. *Ivey Business Journal*, *75*(5), 26–28.

Goodwill. (2016). *Gig Economy Provides New Options to "Liquid Workforce"*. Retrieved from http://www.goodwill.org/blog/newsupdates/gigeconomyprovidesnewoptionstoliquidworkforce

Grohowski, R., McGoff, C., Vogel, D., Martz, B., & Nunamaker, J. (1990). Implementing electronic meeting systems at IBM: Lessons learned and success factors. *Management Information Systems Quarterly*, *14*(4), 369–383. doi:10.2307/249785

Hoang, M. H. (2015). New models of human resource management in China and India. *Asia Pacific Business Review*, *21*(4), 560–562. doi:10.1080/13602381.2014.976976

Kramer, L. W. (2010). Generational diversity. *Dimensions of Critical Care Nursing*, *29*(3), 125–128. doi:10.1097/DCC.0b013e3181d24ba9 PMID:20395729

La Rosa, M. (2016). Interview with Michael Rosemann on The Role of Business Process Management in Modern Organizations. *Business & Information Systems Engineering*, *58*(1), 89–91. doi:10.1007/s12599-015-0419-8

McFarland, W., & Jestaz, D. (2016). Talent development in the digital age: A neuroscience perspective. *Industrial and Commercial Training*, *48*(2), 74–79. doi:10.1108/ICT-09-2015-0062

McGuigan, E. (2016). *Liquid workforce: building the workforce for today's digital demands.* Telegraph. Retrieved from http://www.telegraph.co.uk/sponsored/business/digitalleaders/futureseries/12161816/liquidworkforce.html

Nair, A., & Gaur, V. (2016, September 10). *ICICI Bank introduces software robotics.* Retrieved from: http://economictimes.indiatimes.com/industry/banking/finance/banking/icici-bank-introduces-software-robotics/articleshow/54258671.cms?utm_source=facebook.com&utm_medium=referral&utm_campaign=ETFBMain

Neuner, J. (2013). *40% of America's workforce will be freelancers by 2020.* Retrieved from http://qz.com/65279/40-of-americas-workforce-will-be-freelancers-by-2020

Pegasystems. (2016). *Liquid Workforces Will Let Businesses Put Customers at the Center the digital age is forcing companies to be more fluid about organizational frameworks.* Retrieved from http://digitalevolution.eiu.com/liquidworkforceswillletbusinessesputcustomersatthecenter

Phillips, J. (2010). Open innovation typology. *International Journal of Innovation Science*, *2*(4), 175–183. doi:10.1260/1757-2223.2.4.175

Power, B. (2014, April 23). How GE Applies Lean Startup Practices. *Harvard Business Review*. Retrieved from https://hbr.org/2014/04/how-ge-applies-lean-startup-practices

PWC. (2011). Millennial at work: Reshaping the workplace. *Price Waterhouse Coopers*. Retrieved from https://www.pwc.com/m1/en/services/consulting/documents/millennials-at-work.pdf

Rittgen, P. (2010, June). Success factors of e-collaboration in business process modeling. In *Proceedings of the International Conference on Advanced Information Systems Engineering* (pp. 24 - 37). doi:10.1007/978-3-642-13094-6_4

Sidibé, M., & Campbell, J. (2015). Reversing a global health workforce crisis. *Bulletin of the World Health Organization*, *93*(1), 3–3. doi:10.2471/BLT.14.151209 PMID:25558098

Simons, M. (2016, April 22). *Transforming the Workforce in the Digital Era*. Retrieved from http://www.cio.co.uk/it-leadership/transforming-workforce-in-digital-era-3638972

Taylor, K. (2013). *6 Top Crowd funding Websites: Which One Is Right for Your Project?* Retrieved from http://www.forbes.com/sites/katetaylor/2013/08/06/6-top-crowdfunding-websites-which-one-is-right-for-your-project/#668be67265f8

Wells, D. (2016, May 9). If you're not tapping into the Liquid Workforce, You're Not Innovating. *Inc.com*. Retrieved from http://www.inc.com/donna-wells/if-you-re-not-tapping-into-the-liquid-workforce-you-re-not-innovating.html

Zaino, G. (2005). Freelance Workers: Individual and Corporate Perspectives. *Stcwdc.org*. Retrieved from http://wdcb.stcwdc.org/wp-content/PDF/Individual_and_Corporate_Perspectives.pdf

KEY TERMS AND DEFINITIONS

Ad-Hoc Training: A customized training that is given only to serve a very specific and unique need.

Crowd Funding: A process in which a large project or venture is funded by raising or pooling money from masses.

Gig Economy: A type of work setting in which positions are mainly temporary in nature and jobs are on contracts that are applicable only for a short-term engagement.

Liquid Project: A platform that work on the idea of facilitating work within society rather than walled offices by providing connectivity to those who look for new projects, inspiration, and insights.

Liquid Workforce: A type of tech-savvy workforce characterized by high flexibility, agility, collaborative spirit, sensitivity to the environmental changes, and inherent ability of adapting itself continuously as per organizational requirements.

Millennial: A term used to refer those people who reached their young adulthood around this millennium that is year 2000.

Predictive Analytics: A type of data mining that predicts occurrence of an event in the future by establishing trends from the past data.

Chapter 2
Crowd Funding:
Organizing Finance the New Way

Surbhi Gosain
Bhai Parmanand Institute of Business Studies, India

ABSTRACT

Financing is an essential element for running any organization without which an organization cannot even be created. With many new organizations coming up and foreign companies entering into the market competition has touched its peak. This is valid even for scenarios where companies have to face a cut-throat competition to get the investors for themselves. New and booming startups have to compete with already establish companies where they are generally less preferred as an investment option by the investors who eventually choose big companies because of their return. Thus, new methods of financing such as crowd funding are making their way into present day business environment. Crowd funding has been positioned as a viable option in line with traditional methods of financing such as venture capitalism and angel investments. Crowdfunding has gain popularity among new and inexperienced entrepreneurs and for getting money to carry out social, cultural and environmental projects. Various crowd funding platforms have come up with different pricing and operational strategies.

INTRODUCTION

India is a rapidly growing economy. The government has taken up several initiatives to promote the economy. Many of such initiatives include "Start - up India", "Digital India" and "Make in India". This had given rise to many entrepreneurs in the

DOI: 10.4018/978-1-5225-3191-3.ch002

country and the number is increasing day - by - day. Many of these entrepreneurs are new into the business world with no experience. This is popularly called start - up culture these days with a large number of startup firms mushrooming. Further, already established business houses or companies are also looking for opportunities to expand and diversify for which they require funds. Finance, therefore, is an important resource for setting up a business and successfully carrying it further i.e. surviving in the market. Finance is required for carrying out all the business activities. Finance directly affects the marketing and human resource in an organization. It plays a vital role in a case of technology - based organizations as lack of funds will have a serious impact on research and development process. The arrangement of funds is an integral part of running a business as money is required either to carry out operational activities or for purchasing assets.

Globalization has further added complexity to this scenario with foreign companies entering the markets. Foreign companies may bring with them foreign money and could pose a competition in obtaining finances. Relatively better economic conditions in countries outside India may result in the scarcity of investors investing in India. In such case, startups will have to fight with well - established organizations to get investors for funding their business. Investment in earlier days was basically from a few persons or institutions. It has been seen that investors are bias towards big companies who are already operating in the industry rather than giving preference to young budding entrepreneurs. This inclination of investors towards big companies may also be because of the reason that some investors invest their money in such companies to get huge returns and these companies are backed by their past performance records. Startups will, therefore, have to put in extra efforts to attract investment in their ideas. Also, investment by a wealthy person or by an organization in a startup involves the risk of controlling their operations. Crowd funding in present scenario emerged as a solution to these issues of financing.

Digitization had given rise to the whole crowd funding process as well. The emergence of online crowd funding platforms has provided the investee with an interface to interact with various people who can be their potential investors. This had saved time and effort which a person usually spent on visiting the office of one single investor. Through these online platforms, a person can approach many potential investors at one go. Crowd funding helps in raising money for social projects as well as business projects. Return on investment in crowd funding may vary from just little rewards like vouchers or free merchandises to equity in the company or share in first profit whenever it may happen. Crowd funding has converted mere financing activity into e - financing which had been possible because of the break - out of internet revolution.

BACKGROUND

Crowd funding is an application of crowd sourcing which is a sourcing model, where individual or organizers use e - platforms to reach others for their ideas and resources.

Pathak & Kaur (2016) see crowd funding as a way to "solicit funds" from others to accomplish their ventures or dream projects through online portals either in the form of monetary donation, exchange for a future product, service or reward. It provides an opportunity to small businesses and start - ups with to increase their social media presence, investment base, and funding prospects (Mollick, 2013). It has the potential to increase entrepreneurship by widening the pool of contributors that go beyond the realms of owners, relatives and venture capitalists (Shivaprasad & Kannadas, 2016).

Crowd funding can be categorized into four major categories as mentioned by Pathak & Kaur (2016); Sarkar (n.d.) -

- Donation and reward based: In this, the individual or organization accepts funds in the form of donations from various contributors. For example, establishing a religious pilgrim or any other charity function.
- Lending based: Here the crowd funds take money from different people and then facilitate loans or microfinance to the needy. "Milaap" is an example of such a crowd funding.
- Equity - based crowd funding: In this mode, the contributor(s) take the share of equity in the project or start up, and share the profit as their return.
- Peer to Peer lending: It's an online mode of crowd funding where the online platform acts as a match - maker and matches lenders/ investors with borrowers/ issuers in order to provide unsecured loans. In this, the interest rate is set by the platform.

To initiate crowd funding, one has to create a profile on any of the available online portals of crowd funding or create one's own and explain their projects to raise funds. Jhaveri and Choski (2014) suggested the steps for initiating a crowd funding appeal -

- Identification of a project / explain the cause behind the project by the initiator.
- Selection of the crowdfunding platform (CFP) that fits well with the project / cause OR starting own platform.
- Introduction of the project / cause to the crowd (prospective fund providers) and convincing the crowd why should the fund providers invest / donate.

- Story telling: the story should be understandable and should be able to make connection with the listeners.
- Updating the progress of the project/ cause on the crowd funding platform (CFP).
- Receiving contribution made by the crowd through crowd funding platform (CFP).
- Start the project.
- Earn returns.
- Reward the crowd through crowd funding platform (CFP).

Angel investors are the individuals who have surplus money with them and are interested in investing that money in upcoming business enterprises. The primary reason for their investment is that they want to help young and budding companies to prosper in the market in return for getting a share in the equity of such companies. They are themselves experienced individuals just like a venture capitalist. They also provide the new entrepreneurs with mentoring and solutions relating to their business. These investors either invest their money independently or as part of a group. These days lots of websites have come up which provides the list of these angel investor and dwell the details about their area of investment.

Angel investing is the practice where high - net worth individuals invests their own time and money in new businesses with the goal of profiting from the company's long - term growth.

An angel investor is a person who provides capital, in the form of debt or equity, from his own funds to a private business owned and operated by someone else who may or may not be a friend or family member (Shane, 2008).

Shane (2008) in his report The Importance of Angel Investing in Financing the Growth of Entrepreneurial Ventures mentioned following types of angel investor -

- **Institutional Investor:** A corporation, financial institution, or other organization that uses money rose from another party to provide capital to a private business owned and operated by someone else.
- **Friends and Family Investor:** An individual who uses his own money to provide capital to a private business owned and operated by a family member, work colleague, friend, or neighbor.
- **Informal Investor:** An individual (not an institution) who uses his own money to provide capital to a private business owned and operated by someone else.

One point which is common in above mentioned types of angel investors is that every angel is an informal investor but every informal investor is not an angel. Another important point that comes from these definitions is that there is heterogeneity among angel investors i.e. they either can be accredited investors or not, some are early - stage capital providers, while others put money late, some are passive or they can actively involve with the companies, some, some take high risk while others seek lower risks, some invest alone while others invest as an organization. These different dimensions affect the range of businesses in which angels invest the organizational arrangements that they employ their investment criteria, their decision - making processes, and a host of other things that make describing business angels quite difficult. (Shane, 2008)

Angel investing is important because it fills the gap between what friends and family can offer and what banks, VCs, and Private Equity (PE) firms are willing to contribute. As the former also provide business knowledge, networks and mentorship to entrepreneurs. Through their investments in start - ups, angel investors contribute to job creation.

Venture capitalists are those investors who invest their money in startups or existing enterprises which are not involved in offering their equity to public to generate finances for the company. The key motive behind their investment is to get higher returns on the amount of money invested by them. They assume high risk by investing in startups with a viewpoint that if such a company succeeds then they will get high returns. Therefore, before putting in their money in a company they make a calculative decision about the strength and weakness of that company and their chance of growth in the market. They are experienced business persons with a variety of resources which they extend to the companies in which they make an investment to make them grow and survive the competition. They are well versed in investing a large sum of money in comparison to angel investors.

Ansari and Mohd (2009) defines Venture capital as "an investment in the form of equity, quasi - equity and sometimes debt, straight or conditional, made in a new or untried technology, or high risk venture, promoted by a technically or professionally qualified entrepreneur, where the venture capitalist expects the enterprise to have a very high growth rate, provides management and business skills to the enterprise, expects medium to long term gains and does not expect any collateral to cover the capital provided".

Venture capital is seen as "an activity by which investors support entrepreneurial talent with finance and business skills to exploit market opportunities and thus obtain long term capital gains". Venture Capital can be defined as independently managed, dedicated pools of capital that focus on equity, or equity - linked investments in privately held, high - growth companies. Hence it can be said that venture capital is an investment in the new or growing business either in the form of equity, quasi

- equity, debt or conditional. Since the investment is made in the new ventures, this increases the risk and hence also termed as high - risk investment. (Rani & Katyal, 2015).

Rani and Katyal (2015) listed some of the features of venture capital -

- Equity Financing
- Managerial assistance
- Rewards through capital gains
- Long term investment
- Investment in young start - up companies

The origin of this concept can be traced in USA World War II, where a set of intermediaries invested in young firms that have the potential for extremely rapid growth. From its beginnings on the East Coast of U.S., venture capital gradually expanded and became an increasingly specialized institution. Eventually the venture capital industry shifted from New York and Boston on the East Coast to Silicon Valley on the West Coast Florida. In India, the venture capital started in 1972 when government examined strategies to promote small and medium enterprises and in 1975, Industrial Financial Corporation of India (IFCI) formally introduced venture capital financing in India with an aim to encourage professionals and technologist to form new industries (Ansari & Mohd, 2009).

MAIN FOCUS OF THE CHAPTER

Objectives of the Chapter

The objective of the chapter is to study the process of crowd funding, the policies and pricing strategies adopted by various crowd funding websites. Also, the objective is to study the difference between crowd funding and other traditional methods of financing.

With an increasing number of companies undertaking new projects and many new start - ups in making, a sort of competition has been created in gaining the attention of the investors. With so many ideas, investors have a variety of projects with a different combination of risk and return to put their money into those projects. This has led to people looking for different options of financing to fuel their organizations. That's where crowd funding has come in to help small businesses to fund their business dreams with funding from the public. This is possible over a large number of crowd funding sites been launched in India. These sites have different policies and pricing strategies and are accessed by normal people who are generally not experienced investors. Therefore, there is a difference in the ideology of the traditional investors and the crowd funding these organizations.

Crowd Funding a Campaign: Where to Start?

A person or an entrepreneur seeking money to initiate its venture can appeal the crowd to invest by using various crowdfunding websites over the internet. Some of the popular crowd funding websites operational in India are impactguru.com, ketto.org, milaap.org, wishberry.in and start51.com along with several others.

The first step in the process of raising funds through crowd funding is to create an account on such crowd funding websites and create a campaign for your project. Campaigns are based on any business idea or on a social welfare concept. The campaign created by fund - seeker i.e. an entrepreneur or any individual in need of money can be assessed by various investors or the general public who might be willing to invest if the project is appealing to them. Funds can be generated for either profit - making organization or non - profit organizations. Crowd funding helps to raise funds even for social entrepreneurs along with profit making enterprises.

The fund - seekers generally make videos about their ideas in which they explain their idea, its benefits and how they plan to achieve their goal. They might also put some photographs to create a greater impact on the minds of the potential investors. Visualization of the idea improves the understanding level of investor. It helps in establishing a connection between investee and investor which is essential to convince investors to put their money in someone else's idea. In the case of crowd funding, it is crucially important that an investor is well associated with the idea as the investee is not in direct contact with potential investors and investor will be investing is money only on the basis of the idea. At times investee also gives his contact details so that an investor can clear his doubt about the idea and is complete assured about his investment. Along with videos and pictures, a fund - seeker generally provides the investor with possible returns or rewards which may be monetary or non - monetary in nature. If it is equity - based crowd funding, then the return will be in form of percentage return on the principal amount. If it is reward - based crowd funding, rewards might be in form of some tangible goods or might be exclusive rights to get the products of that new venture at concessional rates.

Each idea is listed on the crowd funding website for a specific short period, varying between days to months, during which it is showcased to the public acting as a potential investor and is available to them so that they can invest in that idea or project if they wish to. After the closure of showcase period, the total amount investment received in favor of that particular project is credited to the account of the investee. Ideas or projects endorsed on such platforms are at first funded by the friends and family of the project initiator and the majority of the funds are collected from the general public who are in support of the idea. These websites provide various services to assist fund - seekers in attracting potential investors and convert them into a successful investment. These websites charge fees depending upon the range of services they provide.

Crowd funding is based on the concept of formation of a community of like - minded investors who have faith in a similar idea and are able and willing to support that idea. Crowd funding is marketing an idea or a concept to attract finances. It integrates marketing function with finance. In other words, it is marketing the financial requirements of the company.

Pricing and Policies

Crowd funding is proving to be a popular method of obtaining finance amongst students and amateur who just have an innovative idea and no network to get funds. They make use of several crowd funding websites available over the internet which can be accessed from anywhere and at any time without affecting the routine work of such fund - seekers for whom it might not be possible to run after private equity. Different crowd funding websites have adopted different policies in case of dealing with the investments.

Ketto: Ketto.org charges highest of Rs. 2000 or 6% of the total amount collected and not the goal amount beside 3% of gateway charges and service tax as applicable from the campaigner and no fees is charged from the donors or investors. Ketto allows the fund - seeker to get the money collected from its campaign once it has ended. Ketto does not transfer money to the account of the campaigner for an on - going campaign. It transfers the money after 14 days from the end of a campaign and once all the documents are provided by the campaigner. Ketto allows disbursement of money to fund - seeker even if the total contributions not reach its goal target. The goal is the target amount with which a campaigner wants to start its venture. It also allows a refund to its donors within 5 days of making a donation and before it has been transferred to the campaigner. There are no time limits or minimum goal levels. Ketto allows only one campaign for one cause. (Ketto, n. d.)

Milaap: Milaap.org has a different range of prices varying with the variety of services offered by them under a package. They have two such packages "do - it - yourself" under which they charge 5% of funds received and "Assist" under which they charge 8% of funds received along with payment gateway charges varying from 1.8% to 2.9% and service tax applicable. In the first package, they provide facilities such as real - time support, campaign manager, multiple fundraisers and daily access to funds while the second package gives additional benefits of features like social media and email toolkit and content creation. There is also a third type of package, namely "Amplify", whose price is decided on basis of needs and demands of the campaigner. It gives further additional advantages of promotion, custom URL, and social media outreach. Milaap allows a campaigner to withdraw the amount raised by him anytime even if the campaign is still on. There is no time - limit or minimum amount to be set as goals. A fundraiser campaign can only be stopped by the campaigner himself. The fun - seeker will get the money even if he fails to meet the goal amount set by him. Milaap permits multiple campaigns for a single cause. (Milaap, n. d.)

Wishberry: Wishberry.in is another popular crowd funding platform functional in India. It is a reward - based crowd funding platform which allows an individual to raise money for creative projects in fields of dance, films, theater, art, design and photography, comic and publishing. It is strictly meant for creative and productive projects and does not support social or community welfare projects (Wishberry, n. d.). Wishberry charges a fixed non - refundable fee of Rs.3500 plus service tax for a defined range of service provided by them such as personalized consultation, global payments and trouble - shooting, backer management tools and E - marketing tools. Wishberry takes 10% of commission from fund - seeker if they are able to achieve their goal amount. Moreover, wishberry provides digital marketing and PR services at a price of Rs.35, 000. Wishberry allows withdrawal of funds only if the fund - seeker is able to achieve its goal. Any amount of money received by a campaigner will be refunded to the donors of such money if the campaigner fails to meet its goal. (Campaign management fees, Wishberry, n. d.)

Impactguru: Impactguru.com charges 5% on the amount raised along with 5% as transaction cost and service tax applicable on it. It offers services such as integration with social media platforms, consultation, fundraising events and relationship management facility. (Impact Guru, n. d.)

Start51: Start51.com is another crowd funding platform following a timed campaign policy. Each campaign is showcased for a particular duration during which it has to raise its target amount. If the campaigner fails to reach its target or could not attract enough contributors for its campaign then he will not be getting any money. A fund - seeker will be allowed to withdraw the crowd - funded money only if he is able to achieve its target otherwise any money collected will be refunded to its donors. The name Start51 is indicative of the time - limit for each campaign i.e. 51 days. It means that each campaign will be given a time of 51 days to attract funds after which it will be automatically closed. Start51 does not endorse charity campaigns or campaign initiated by individuals seeking money for their personal use. This is only meant for financing the projects which would lead to some productive outcome in the social, cultural, environmental or entrepreneurial field. It does not support equity - based crowd funding and is only based on reward - based funding. Start51 charges 5% on the contributions received plus a processing fee up to 2%. (Start51, n. d.)

Each of this crowd funding websites requires the campaigner or fund - seeker to set a specific target of money which he wants to start his venture or project. Further, they advise their users to keep their campaigns updated by posting about the progress of their project which keeps the interest of donors alive in their campaign and helps increase the magnitude of contributions. They facilitate the services which enable a channel of communication between the fund - seekers and their potential investors who might also be their future customer in case of a product or service - based startups.

Crowd Funding, Venture Capitalist and Angel Investor: How They Differ?

Venture capitalism and angel investment are the traditional methods of financing whereas crowd funding is newest and modern-day financing option adopted by many young entrepreneurs. The traditional methods are very much different from crowd funding in terms of expectations and perception of investors.

Crowd funding is mainly used by small or medium scale enterprises that are in need of the initial investment to start their organization. Large enterprises or companies do not take the route of crowd funding as such companies have high demands of finance which cannot be met using this method of financing an organization. Most of the large - scale organization are either listed on stock exchange and procure their funds by the issue of shares or debentures or obtain funds through private equity, as in the case of private companies, by the methods of venture capitalism or loans. Thus, crowd funding is capable of generating a pool of funds for starting a venture and carrying out initial operations but organizations generally go for more sophisticated ways of financing for development stages as the demand of financing swells to a level where it is unlikely to meet a large number of small investors. Angel investors invest in a startup. They provide money to a company looking for funds to venture out in the market and have a business model which is capable of generating returns. VC invests in already running companies with well - established business operations and with a scope of further growth.

Angel investors look for ownership or equity holding in exchange for their investment while Venture capitalism wants a rate of return matching with the level of risk. Crowd funding has no such issues involved as the individual investors have no such expectations of returns. There are small Rewards in crowd funding. (Venture-Capital-Investment.co.uk., n. d.)

Venture capitalist and Angel investors lead to dilution of ownership in an organization as they hold a position on the board of the company. Venture capitalist keeps a watch on every move to ensure the safety of his investment while angel investor acquires stakes in the company for making the investment. In other words, they have the authority to make decisions and hold a position where they can influence the decision of the company (Venture-Capital-Investment.co.uk., n. d.). Crowd funding does not dilute the equity of the company except in a case of equity - based crowd funding and hence have no interference in working of the company.

Crowd funding method of generating finances can be used for funding profit based organization or non - profit organization but venture capitalist only invests in those ideas which have the capacity to create returns. Crowd funding can be helpful in initiating a social, cultural or environmental project which may be beneficial for the whole society. Thus, these types of project are more popular on crowd funding platforms as society is pooling in money for their own future good. The primary motive of a venture capitalist is to secure his return, unlike crowd funding. Individual investments through crowd funding are mostly independent of the expectations of return. Although equity - based crowd funding has started to gain acceptance but it is still not well established as it involves lots of legal issues and problems. On the other hand, private equity methods cannot be used for funding any project other than a profit - making venture.

A venture capitalist is more likely to fund an organization after undertaking a systematic analysis of the market conditions and the assessing the strengths and weakness of the organization. They follow a tendency to invest in those ventures that are similar to or belongs the area of their existing investments. While crowd funding involves funding on the basis of an individual's association with the idea or concept proposed by the organization. Such an association may be in an emotional context or logical context. Similarly, angel investors also make a decision regarding investing in a particular venture only after detailed inspection of the office premises, analyzing the workforce indulged in working of the project and completely satisfying themselves that the venture has a strong growth potential.

Venture capitalists and angel investors are approached by the organizations that tend to have an extensive network of people who provide information about possible investors, unlike crowd funding where no connections are present. Networking is of uttermost importance in a case of venture capitalism and angel investments. An entrepreneur needs to have connections with people who can provide them with the details of such high - profile investors and fix a meeting with them.

Venture capitalist and Angel investors are the platforms which are only meant to arrange funds for the project but crowd funding provides a platform where the fund - seeker can get to know about the future of their idea and whether their product will be accepted amongst the public. They get a chance to know about their future in advance. Crowd funding helps in marketing of an idea. It involves extensive usage of social media platforms such as Facebook, Twitter or YouTube.

Venture capitalists prefer to invest in those companies which have been an outcome of a successful project. In other words, they usually back those organizations which either have some past performance record or are based on an idea which was proven success on small scale and needs fund for its execution on large scale. Angel Investor seeks to invest in those companies which display a growth potential and they can benefit by selling their stakes in such companies when they have achieved a higher value in the market. They generally hesitate in funding a completely out - of - the - league idea. They judge the project by the entrepreneur. Thus, an amateur may find it difficult to get finance from a venture capitalist. This ensures the venture capitalist and angel investor that whether the fund seeking individual or organization is capable enough to handle the risk and carry his business successfully. These investors are experienced business persons who take their investment decision after analyzing all the aspect of a venture. While on the other hand, the crowd who funds the project has no knowledge about the abilities of the individual seeking their money. They are not the experienced investors who invest their money by taking calculative decisions. (Mollick, 2013)

Venture capitalism and angel investment require face - to - face meetings with the organizations seeking investments which involve a lot of time and efforts. Crowd funding involves meeting with a number of potential investors through a visual presentation over the internet which is relatively less time - consuming.

Impact on Organization: How Does Crowd Funding Make A Difference?

Crowd funding has paved the road for innovation. It has led to the IT - based mobilization of new ideas and solutions to the organization's problems. Therefore, it makes organization more technology - savvy.

Organizations have to present their ideas to a wider audience and not just for few people in a boardroom. The ideas and concept behind the project should have a wider reach to impact multiple investors and stakeholders. Crowd funding ensures a better relationship between organizations and their stakeholders. Crowd funding has also prompted business organizations to change their business models. Business models define how a business organization functions and optimize its resources. Crowd funding has introduced a crowd - based business model wherein crowd has the involvement in everything from product design and software development to financing.

Crowd funding has forced the companies to change the orientation of their product designs, production process and marketing activities to tap the funds from the crowd. The success of an organization is dependent on the pool of crowd and its quality. Quality here means that whether the people are genuinely interested in investing in something new and innovative or just do it for fun purpose. A genuinely interested pool of crowd - funders can lead to the on - time completion of the production process and put pressure on the company to meet its deadline as they follow the campaign even after it's completed. They are keen on updates by the company.

Crowd funding requires an organization to showcase its project management capabilities. For this purpose, organizations need to create an effective project proposal along with how they plan to accomplish their objectives. This helps an organization to get a clear vision about their path ahead. Enterprises have to provide the crowd with a clearly defined operational plan and its end results to create a large pool of crowd. This requires having an efficient team of personnel's who knows the project in - and - out and is capable of presenting it to crowds. They should be well versed in the queries of the crowd and carrying out an interaction on a one - to - one basis. An effective and efficient team of personnel with all the required skills and capabilities is an asset for the organization. So, crowd funding act as a testing tool for judging the performance of the workforce.

Crowd funding platforms provide a base to launch first models of a product or service and at the same time puts an onus on the enterprises to come up with better rewards or additional features based on the feedback received from previous models. This ultimately helps the enterprise in the future development of its product, production and personnel.

Crowd funding allows a company to have a low inventory cost. It provides great access to working capital. People provide funds for a particular project which is used as fuel for the production process. Unauthorized the burden of inventory cost. Organizations have to be more responsible from beginning to end. Digitization gives a competitive edge to the company. It opens access to new resources. Crowd funding has opened gates for more finances.

Crowd funding also helps the funder - seeker, be it a company or new entrepreneurs, to access the preference of the target audience. If a venture is successful in reaching its target goal, then it is an indicator that they are heading in right direction. Social media marketing has significantly reduced marketing cost.

Digitizing Finance

Crowd funding organization with generating funds via the internet. It has brought a whole new era of digital culture in the organization. It has digitized the very basic need of organization - money to put its business concept into practice. It involves using internet - based platforms and various social media platforms to connect investor and investee. In present times, the internet has made its reach to every nook - and - corner. It had provided connectivity to more number of people. Society, today, is well equipped with computers and technology. We are living in a computerized era. Hyper - connectivity has fueled crowd - based tools where people not only access the resources but also contribute to them.

Digitization ensures more transparency, security and integrity. Development of high - end software has enabled the investors and investee handles their projects on their own. It helps to create a large database for an organization's future use. The rise of social media has also lead to the emergence of crowd funding. People used to connect to each other using social media platforms. Crowd funding has channelized this into a well - structured software based platform which has added various security and control features to it.

Case of Formlabs: A Successful Crowd Funding Campaign

Formlabs is a company engaged in manufacturing of 3D printers with an aim of making it accessible and used by engineers, designers or artists. Formlabs was founded by Maxim Lobovsky, Natan Linder, and David Cranor, a group of engineers and designers from MIT Media Lab and Center for Bits and Atoms, in September 2011. Formlabs printers are the top selling 3D printers in stereo lithography worldwide. The company has also introduced its high - quality materials for 3D printing. The company has over 190 employees across North America, Europe, and Asia. Formlabs has launched its flagship product, Form 2 3D printer that uses stereo lithography (SLA) to create high - resolution objects. Its objective is to create professional desktop 3D printing.

Formlabs was started in 2011 with an initial seed investment from various persons including Mitch Kapor, Joi Ito, and Eric Schmidt's Innovation Endeavors. The company opted for crowd funding for its full production endeavor. They selected Kickstarter.com to launch their campaign. They launch their campaign on September 26, 2012 for a period of 30 days i.e. up to October 26, 2012. Their target goal was $100,000. They were able to raise around $2.95 million with a number of 2,068 backers for their project during the tenure of their campaign. They crossed a mark of $100,000 within 2.5 hours of going public with their campaign. This was termed as the most successful campaign in the history of the Kickstarter.

Formlabs had crowd funded the production of their first 3D printer, Form 1, after carrying out the complete research and development process to find out a solution for building an affordable 3D printer. They adopted a reward - based funding where the packages varied from a mere T - shirts and 3D printed Gyrocube with the company's logo to the 3D printer itself. Those who pledge $2,299 or more were rewarded with the Form 1 printer, resin, and the form finish kit.

On November 1, 2012, Formlabs fulfill its first reward promised by shipping the miniature printer model to all those who had pledged $5 or more. On November 21, 2012, the company launched its own "Form Store" for its customer to pre - order the printer. In December 2013, Formlabs completed the shipment of the first lot of the actual Form 1 3D printers to the people who had pledged $2,299 or more. On October 24, 2013, company went for Series - A funding and was able to raise an amount of $19 million from a number of venture capitalist and angel investors. Further, the company had introduced new and improved versions of its 3D printers, namely, 'Form 1+' and 'Form 2' on June 10, 2014 and September 22, 2015 respectively. The company had also launched various 'materials' which are nothing but the ink for its 3D printers called resins and a 'Preform' which is a software package.

On August 4, 2016, Company went for series B funding and raised $ 35 million taking total investment to $ 55 million. The money was invested by venture capitalists.

SOLUTIONS AND RECOMMENDATIONS

Although crowd funding has been adopted by many as a way of initiating their ventures without losing control over their business and idea but it still has to deal with certain complexities to beat the traditional methods of financing.

Crowd funding is seen as a whole new industry which can provide various employment opportunities. The emergence of crowd funding platforms had given employment to a number of individuals and has created new job profiles such as 'campaign - handlers', 'social - media advisors' and much more.

Equity - based crowd funding is more preferred by the structured investors who prefer small returns or holding in the company. Crowd funding mechanism requires proper channelizing to safeguard the rights and interest of investors, especially in the case of equity - based crowd funding. This involves the formulation of policies and even organization acts to monitor the flow of funds under this mechanism. JOBS Act in the US had put some rules in place to deal with this problem. While in India equity - based crowd funding has been stated as an unauthorized way of raising money by securities and Exchange board of India (SEBI). Such a practice is seen as violation of Companies Act and Securities Act. Any company raising money by issuing shares can do so only recognized stock exchange and not on any crowd funding platforms. This has been criticized by the private investors. Thus, there is much scope for proper regulation to resolve this issue in future.

Crowd funding has no longer been limited to start - up only but can be helpful for fully functional corporate business houses. Providing education to the fund - seekers as well as to the potential investors is important to ensure a robust campaign and healthy investment option for investors. There is a lack of awareness and confidence among people on using crowd funding as an investment option.

CONCLUSION

Crowd funding is a breakthrough innovation in the field of financing an organization enabling the common middle class to invest in companies. Crowd funding is the practice of funding a project, be it social or business, using a small amount of funds from a large number of people. Crowd funding has given an option to a number of small and young entrepreneurs to kick start their ideas and ventures with an easy financing facility. Other sources of financing a business may include Angel investments, Venture capital or raising loans from the bank. All these options require an investee to approach the investors with a radically innovative idea which can be converted into a product or service in real - time and is capable of earning record profits along with assuring the investors the required rate of return on their investments. Under these financing options, entrepreneurs have to promise a high rate of return to the investors. Thus, these options pose a difficulty for a fresh talent to gather money for his idea. This is simply because of the reason that it may take time to convert one's idea into profit and thereby causing a delay in creating the return on investment. Crowd funding allows any person to be an investor it may not be necessary that the person must possess a fortune of money. Crowd funding also allows an investor to be emotionally connected with the idea he is supporting. Angel investment or venture capitalism at times limits the freedom of the entrepreneur but crowd funding provides an entrepreneur with its liberation to work on its creative terms. Although, Reward - based crowd funding is more popular in India but equity based crowd funding is also gaining momentum.

In conclusion, crowd funding emerged as the potential investment strategy that can minimize the risk and maximize the field of innovation and experimentation which leads to many new promising inventions and discoveries that are very important to boost the economy of a developing country like ours.

REFERENCES

Ansari, A., & Mohd., A. (2009). Role of venture capital in spurring innovation and entrepreneurship. Asia - Pacific Business Review, (1), 114-119.

Wishberry. (n. d.). Campaign Management Fees. Retrieved January 3, 2017, from https://www.wishberry.in/pricing/#/pricing

Impact Guru. (n. d.). Retrieved January 3, 2017, from https://www.impactguru.com

Jhaveri, H., & Choski, A. (2014). Crowd funding at India: a study of Indian online crowd funding platforms. Retrieved 30 January, 2017 from https://www.researchgate.net/publication/272984988_CROWDFUNDING_AT_INDIA_A_STUDY_OF_INDIAN_ONLINE_CROWDFUNDING_PLATFORMS

Ketto. (n. d.). Retrieved January 3, 2017, from www.ketto.org/support/faqs.php

Milaap. (n. d.). Retrieved January 3, 2017, from https://milaap.org

Mollick, E. (2013). The Dynamics of Crowd funding: An Exploratory Study. *Journal of Business Venturing*, *29*(1), 1–16. doi:10.1016/j.jbusvent.2013.06.005

Mollick, E. (2013). Step away from the crowd? Crowd funding, Venture Capital and the Selection of Entrepreneurs. Retrieved 29 January, 2017 from ssrn.com/abstract=2239204

Pathak, S., & Kaur, P. (2016). Crowd funding: an innovation in the age of digital marketing. International Journal for Innovations in Engineering Management and Technology, 1(1), 26 - 34.

Rani, P., & Katyal, H. (2015). Venture capital in India: sector - wise analysis. International Journal of Research & Development in Technology and Management Science, 21(6), 72 - 80.

Sarkar, A. (n. d.). Crowd funding in India: Issues and Challenges. Retrieved 29 January, 2017, from http://ssrn.com/abstract=2739008

Shane, S. (2008). The Importance of Angel Investing in Financing the Growth of Entrepreneurial Ventures (Rep. No. 331).

Shivaprasad, G., & Kannadas, S. (2016). Is Endowment of Crowd Funding a Supplant of Traditional Source? International Journal of Scientific Research, 5(2), 138 - 141.

Start51. (n. d.). Retrieved January 3, 2017, from http://www.start51.com/help

Venture-Capital-Investment.co.uk. (n. d.). Venture capital investment vs. angel investment. Retrieved January 3, 2017, from http://www.venture-capital-investment.co.uk/venture-capital-investment-vs-angel-investment.php

Wishberry. (n. d.). Retrieved January 3, 2017, from https://www.wishberry.in/rules

KEY TERMS AND DEFINITIONS

Angel Investor: A wealthy person who invests in businesses in exchange of return on his investment.

Crowd: A group of people who share a common interest in subject and form a community to support it.

Crowd Funding: The process of collecting for a project, venture or to start a business from crowd.

E- Financing: A way of generating money for business using internet technology.

Equity: A stock of the company reflecting the ownership interest in that company.

Returns: Interest or any benefit received by the investor on his investment in a company.

Venture Capitalist: A person who invests in start-ups and small business but have no means of entering into share market.

Chapter 3
Crowd Shaping:
A Morphological Innovation

Vishwas Gupta
Lovely Professional University, India

Indranil Bose
University of Bolton, UAE

ABSTRACT

Welcome in today's business where change is the only constant. The companies continue to invent new products and technologies. The innovative and improved technologies transform the industries and companies and change lives. But it involves so many challenges and difficulties. Still they are pursuing for better products to cater the future needs of customers. At IBM Smarter Cities Technology Center, a. team of engineers was working on restructuring the city road system by using the mobile phone data. Almost 2.5 billion call records from five million mobile users in Ivory Coast were used to give a final touch to this operation. The data was used to improve the public transport service in the city. Once the suggestions are executed, the Ivory Coast would be only city to witness the world's first 'Crowd shaped' road system. A road system wisely and beautifully re - engineered with the help of information collected through the data of its commuters. In the chapter we will try to know what is crowd shaping and how is it affecting the life of a retail shopper in this technical savvy world?

CROWDSHAPING: AN INTRODUCTION

At the IBM Smarter Cities Technology Centre in Dublin, a team of engineers were working on restructuring the city road system by using the mobile phone data. Almost 2.5 billion call records from five million mobile users in Ivory Coast were

DOI: 10.4018/978-1-5225-3191-3.ch003

used to give a final touch to this operation. (The personal details of the data were not revealed but breaching of privacy issues was raised in later part of the operation). The data was used to improve the public transport services in the city.

"The data provide us some of the very useful information like the way people roam around the roads in city, about the various congestion points, the traffic peaks at the beginning and at the end of the working day, the total traffic on a particular destination at a given time". They have developed an efficient and robust computer model which is able to study the billions of data details and recognized the points where some alterations are possible to modify the bus routes and their schedules to cut average travel time of commuters by almost ten percent.

Once the suggestions are executed, the Ivory Coast would be only city to witness the world's first 'Crowd shaped' road system. A road system wisely and beautifully re - engineered with the help of information collected through the data of its commuters. Now the same team is working on parallel project for the city of Dublin, using GPS data of cell phone used by commuters.

Still the real potential of crowd shaping has not been retrieved, when the data can be used to reshape the services in real - time. So far past data has been accessed and researched to make some modification. But in 30 years of span we will have the system that will be able to process billions of GPS mobile data as received and amend the traffic routes and schedules on a minute - to minute basis. Also, the others services will be benefitted. Crowd shaped medical facilities, crowd shaped police patrolling, crowd shaped manufacturing products and crowd shaped retailing services. That stage of crowd shaping would bring the revolutionary benefits for the people in real.

The in-store technology enables the shoppers to notice crowd shaped advertising on digital screens where the shoppers will be able to find precisely his choice of t - shirt with just one thought and it will be clicked on screens showing the exact location and other details of the product. You can crowd shape the in - store music or music at a party aligned automatically with the playlist of your smart phone.

THE EVOLUTION OF CROWDSHAPING

The traditional marketing and product development strategies are based on the information extracted from a data retrieved from a crowd of people or customers since last so many years. The legacy is going on and it is still considered the best way to improve the products and services. But the methods of acquiring the data have been transformed. Big Data, Cloud Computing, The Internet of Things, Real Time Data, and the various disruptive technologies have enabled us for tailor made future where we would be able to mold the required technology and product as our own preferences and convenience. Several companies have harnessed the information collected from the network of individuals in the form of crowd sourcing. The crowd sourcing is more authentic and better solution provider technique in comparison to traditional problem-solving approaches. In fact, the crowd shaping is resultant of sum of the various activities, arising in form of outcome of customizing data at various stages. It can be shown as –

$$CSP = \sum CS, OS, OI, UI, OS$$

Where, CSP is Crowd Shaping,

CS is Crowd Sourcing,

OS is Out Sourcing,

OI is Open Innovation and

OS is Open Source.

The traces of evolution of the crowd sourcing are available since 18th century in Britain (Halder, 2014). In 18th and 19th century the crowd sourcing was used only for scientific invention but in 20th century other areas like arts, commerce and especially in product development it has been widely used (Halder, 2014). The application of crowd sourcing has been utilized in the areas of travelling (Blanc & Figliozzi, 2016), Online marketing (Sukaini, 2016) and fundraising (Mariani, Annunziata, Nacchia & Vastola, 2016) too.

In a dance party in Texas in 2006, each person of the gathering had given a bracelet. Only condition for the dance party was not to take off the bracelet. The bracelet was one of the crowd shaping inventions to measure the heat beat, movement and other physiological measures of dancers. On the basis of responses, the music, lights, food and drinks were decided during the party duration as per the preferences of users.

The term crowd shaping coined as 2014: brainchild of crowd sourcing by David Matin, head of Trends and Insight. It means to gather the real - time data from a gathering of crowd from a defined area in order to provide the tailor-made product or service as per expectations and feedback given.

The process of crowd shaping consists of following steps: Business listing verification, Search relevance, Product/service categorization, Content generation and Custom solutions. There are two types of crowd shaped services. Small crowd shaping, in which the real - time data is used to modify the current products or services like a music concert, discotheque and restaurant where the customers are

Figure 1. Components of crowd shaping

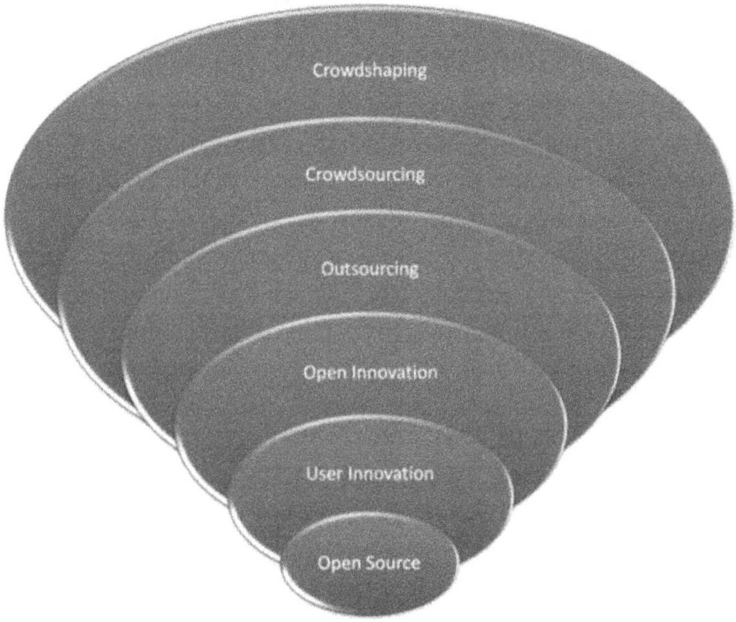

Figure 2. Evolution of crowd shaping

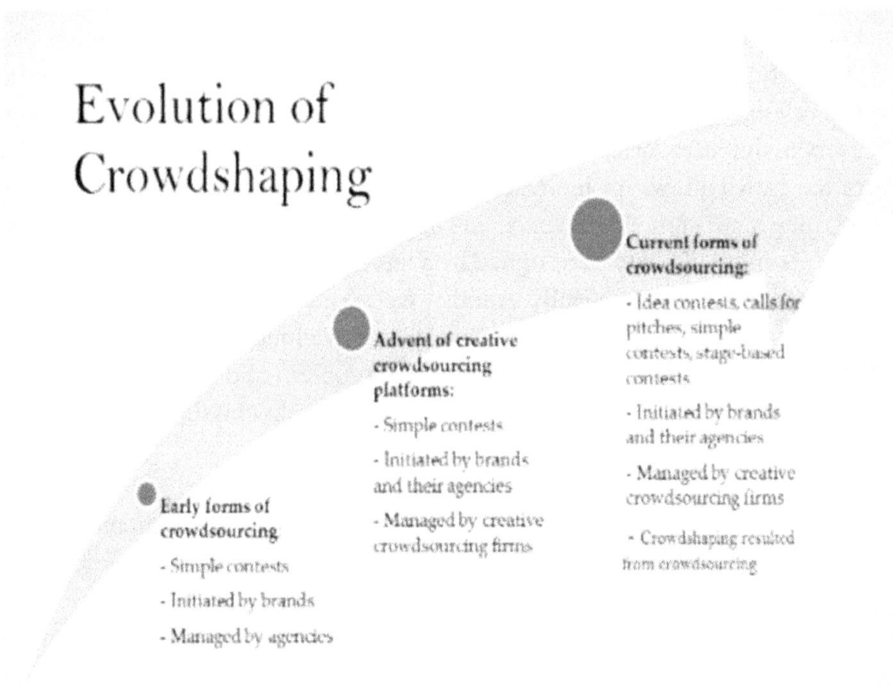

easily monitored and their preference can easily be noticed and implemented. Real - time shaping (and reshaping) of a service around the preferences of the people in an office, a restaurant, on a plane: any space right now. Consumers might not care about the tech that makes CROWDSHAPING possible, but they will care about increased relevance via the accommodation of their own preferences and a novel mode of discovery via the newly experienced preferences of their peers. See how CheckinDJ (details are given below) delivers both.

While the big crowd shaping requires the services of the big data and the information is required on a very large scale based on the large number of the consumers. Services intelligently reshaped by the aggregated data on the preferences or behavior of large numbers of consumers. It is one next step for good, old - fashioned crowd sourcing. Consumers get a more functional, efficient service shaped not by opinions of the crowd, but around the way people really behave. And – unlike with many existing crowd sourced solutions – they get it effortlessly, via passive sharing of their data. That's what IBM promised when it CROWDSHAPED bus route improvements in Africa (details are given below).

EXTRACTION OF DATA FOR CROWDSHAPING

1. **Classification:** Classification can be performed within crowd sourced data. For instance, users can categorize documents or can assign labels, also known as classes (i.e., tags) to documents manually. This approach has been successfully tested in various domains. Examples include, but are not limited to: Yahoo! Directory, social bookmarking sites, and social news. In Digg, users assign categories and tags to submitted links, i.e., diggs. The wisdom of the crowd allows for more accurate categorization than any other machine learning algorithm. For instance, in Digg (Agarwal et al. 2008), more relevant tags are more likely to be assigned by a large percentage of users to an article. This reduces noise drastically. Another well- known example is when humans help to solve a complicated problem where machines fail. CAPTCHAs and the technology behind them help solve the problem of digitizing handwritten text (Von Ahn, 2007). In CAPTCHAs, humans help classify text by labeling images of text that need to be digitized.
2. **Clustering:** Clustering can also be performed using crowd sourced data. In general, many recent social networking sites seek aid from humans in order to create categories. For instance, on Balatarin, a Persian social news site, users automatically create categories. Users are allowed to choose one of these categories for their submitted material. This can be viewed as an instance of automatic clustering of submitted articles. Another example is the case of

Twitter, where users form clusters (known as trends or trending topics) by assigning hash tags to their tweets. This facilitates fast retrieval when searching for tweets and again, adaptively clusters streaming tweets.

3. **Semi - Supervised Learning:** In semi – supervised learning, a learning algorithm is given a subset of labeled data and another subset consisting of unlabeled ones. It is then required to label the unlabeled set using the information acquired from the labeled set. Similar to previous tasks, semi - supervised learning can also be performed with crowd sourcing. Consider the example of Amazon Mechanical Turk. As mentioned in their website, Amazon Mechanical Turk is a marketplace for work that requires human intelligence. The Mechanical Turk service gives businesses access to a diverse, on - demand, scalable workforce and gives workers a selection of thousands of tasks to complete whenever it's convenient. Amazon Mechanical Turkish based on the idea that there are still many things that human beings can do much more effectively than computers, such as identifying objects in a photo or video, performing data de - duplication, transcribing audio recordings, or researching data details. Traditionally, tasks like this have been accomplished by hiring a large temporary workforce (which is time consuming, expensive and difficult to scale) or have gone undone. In Mechanical Turks, humans are given examples of how a given task can be performed correctly and then asked to generalize and solve the same task for some other instance. This is a well - practiced labeling technique for complex data labeling tasks in the data mining field (Sorokin and Forsyth 2008).
4. **Validation:** Similarly, humans can help validate the results of clustering, classification or semi supervised learning. The validation task is performed as follows –
    a. The learning task (classification, clustering or semi - supervised) is performed using an automated technique.
    b. The same task (probably on a smaller scale) is performed by humans (crowd sourcing).
    c. The results are compared with the automated method's outcome and the accuracy is calculated.

This gives an estimate of the overall accuracy of the automated method on a larger scale. In Agarwal et al. (2008), authors propose a method for identifying influential bloggers in the blogosphere. They evaluate their automatic technique by comparing their results to the crowd sourced data generated on Digg. They assume that the number of diggs humans have assigned to the posts submitted by influential bloggers should be higher and based on this assumption validate their findings.

5. **Sampling:** One of the complex tasks in data mining and machine learning is sampling. Due to the scale of current datasets, it is required to obtain samples

with sufficient information so that the hypotheses devised from the sample information can be easily generalized to larger datasets. However, the question that needs to be answered here is how the sampling distribution should be selected so that the information obtained from the sample is maximized. Humans have proven to be credible information samplers, and it turns out the criteria that they aim to maximize results in maximally informative samples (Von Ahn 2007).

RECENT CROWDSHAPING DEVELOPMENTS

1. **Mychiatry:** Each mankind strives from fulfilling the basic needs to the stage of self - actualization. The crowd shaping is working as an aided tool for such kind of 'Personal Psychiatry'. Here the technology is being used as a problem-solving technique to develop the customized product in form of watches, headgear, Google Glass and wrist band. Such crowd shaped product help to improve the mental, physical and emotional health by reporting the real-time data about heartbeat, glucose level, blood pressure and other essential information which make the customer well aware about his or her health. The Nike already has started to use this technology in its fitness products also various app on smart phones are working as a source of crowd shaped products.
2. **Upgradia:** This is the era of technology which is reverse engineered, each company and customers are obsessed with the latest technological upgraded product. The product life cycle of each product has shortened (Chung & Wee, 2011). Very soon the products come in the range of inferior product (Hoegg, Alba & Dahl, 2010) and every company is launching a better modified product each after 6 months. The crowd shaping is like a boon for the companies which are always ready to fetch the endless consumer desires for the next better range of products. Microsoft, Apple, Samsung and almost each company today is engaged with the constant digital innovation.
3. **The Age of Impatience:** This is a smart era, where smart people are endowed with 201smart technologies with the help of smart phones in search of smart products. By the end of 2016 there will be 2.1 billion smart phone users in the world (Statista, 2016) and it may cross 5 billion by the year of 2019. It means one out each 7 persons in the world is using smart phones, which has transformed the personal as well as social life of a person. The buying pattern of customers has also changed. Now the people are well informed about the product and the competitor products too. The crowd sourcing has helped a lot to

the companies in the form of Big Data about the preferences and expectations of the customers now the companies are also well informed and ready with the latest product.

4. **The End of Anonymity:** Digital Footprint - You are no more unknown to the companies today. Call it breach of privacy or thanks to crowd sourcing, your every movement and action is being monitored today by the corporate companies (Savitz, 2012). IBM can easily trace out your personality by decrypting just 200 of your tweets and the number of tweets will reduce in future to just decode your desires (Bernie, 2012). The ethical hackers and the IT and the cyber security professional are going to be the kingmaker of the corporate life in future (ISACA, 2013). The Big data in the form of breaching of privacy, and threat to cyber security can be a big question which raises the questions about the rights of consumers. But crowd shaped business is generally done after consent of the target users only. People are ready today to accept this reality of market. 86% of US online users have reduced their online activities and only 37% still believes that no one can breach their privacy (Lee Rainie, 2013). While in India this threat is the biggest hindrance in the step of complete digitalization of all types of payment initiated by and promoted by Indian government recently.
5. **Permission Marketing:** The digital marketing alone today is not enough to win the customer's heart and faith. We require interactive digital marketing which facilitates the two-way communication system between the customer and seller (Kumar, Zhang & Luo, 2014). The users must be sure about safety and security of data. Permission is taken from the user prior to use his or her personal data as a source for crowd shaped products.

KEY OPPORTUNITIES AND CHALLENGES IN THE CROWDSHAPING

The business of extracting the information from the crowd is not so easy. The companies have to face the questions like where to use crowd? How to use crowd? Which crowd has to be studied? How to support the required technology like SQL or some other? How to devise a system? Whether the quality of data be achieved?

Several studies have proved that the collection of real-time data from the crowd is a challenging task -

- 57% of consumers are willing to share additional personal information, such as their location, top five Facebook friends' names and information about family members, in return for financial rewards or better service. (Parkes, 2013)

- 82% of global consumers believe that companies collect too much information on consumers. (Adobe, June 2013)
- 86% of US internet users have attempted to remove or mask their online activities, despite only 37% believing it is possible to be completely anonymous online. (Rainie, 2013).
- 93% of email users believe that users should be able to opt - out if they don't want the content of their emails to be scanned in order to target ads. (Microsoft, 2013).

The Internet of Things will add USD 1.9 trillion dollars of economic value to the global economy in 2020. In 2009, there were 2.5 billion connected devices; most of these were personal devices such as cell phones and PCs. In 2020, there will be up to 30 billion connected devices, most of which will be products. (Stamford, 2013)

The crowd shaping is not limited to only for development of better products and services but it also cater for the human - computer interaction (HCI), artificial intelligence (AI), machine learning (MI), information retrieval (IR), crowd management (CM), and for the business statistics (BS) too.

In a study of one third of the global consumer class almost 2.5 billion consumers were studied and the data has revealed that 78% of the customers love shopping, 92% have a desire for responsible consumption while only 58% believe in branded products (BBMG, 2013). The crowd shaping helps us to solve our problems and needs in a smarter way. Also, it can be applicable for corporate social responsible way of business.

The smart phone has made every user multitasked. Every user wants to be connected all the time. Checking the mobile phone every 25 min on an average, an uncontrollable habit of Facebook updation, again and again checking the high - frequency email inbox, sharing and clicking the Instagram moments make a user multitasked. There is digital tsunami of data on the internet for both the users as well as for the companies.

The crowd shaping provides another massive opportunity for the companies to focus on the highly stressed, time starved users who needs some better products or services for a better lifestyle. Several healthcare products, fitness products (Both physical apparatus and edible products) have been developed to provide relief to users from the over stressed life.

One of the major challenges for crowd shaping is the task management - which crowd should be invited? Whether that is the right set of gathering to get the required answers? What should be the right demography of people? Whether they should be offered any payment or not? Will it lead to any biased data? And what is the authenticity of the data retrieved here?

Another major challenge is the human computer interaction in form of payment interface, interaction design and mode of communication. Quality control during crowd shaping is another challenge. The trust and reliability of data, the spam detection and consensus labeling are also the challenges for effective crowd shaping.

LEGAL AND ETHICAL ASPECTS INVOLVED IN PRIVACY, SECURITY, AND DATA PROTECTION ISSUES IN CROWDSHAPING

Imagine you get up in morning, check Facebook and find that your status has been automatically updated. Facebook claims that it can now precisely foresee your relationship status by analysing trends and happenings going on in your life. Several financial companies are on a spy spree and they scout each of your social activity on various social networking sites and contact you with the customized investment plan.

The crowd shaping products are generated on the basis of crowd voting – if the product has been developed on the basis of preference, ratings or feedback given to the company in online or offline mode. The crowd creation, where the people create, design, manufacture a product of their choice in a contest held by a company, the crowd wisdom, the people are given some set or problems to solve. And the final crowd funding where people are motivated to donate for various charitable activities.

But there are lot of apprehensions about crowd shaping activities. The target users must be well informed that they are being studied and each and every online activity is being monitored.

Despite enormous scope, the crowd shaping has several limitations too. First the active participation of a crowd, next the authenticity and unbiased information extracted from the crowd.

SCOPE AND BENEFIT OF CROWDSHAPING

The eventual scope of crowd shaping lies in developing the customers and not the products. There are four times more chances that a customer may switch to a competitor's product if he is not satisfied with the service related issues rather than price or product related issues (Zeithamal, 2002). The crowd shaping focuses on the customers experiences. With the help of the profile of users, data and the preferences through social media and other online platform the crowd shaping has helped in the development of more user-friendly technology and also in the form of higher

Table 1. The pros and cons of crowd shaping

Pros	Cons
Potential positive publicity (e.g., Doritos)	Publicity is not always a positive thing (e.g., Netflix Prize)
Create a connection with consumers	Public not always trustworthy, generally cannot indemnify
Give something back to consumers	Consumers may think you could give more, liability risk may outweigh potential benefit
Inexpensive method of acquiring new ideas/content	No verification submissions are non - infringing, costs/burden associated with clearing infringement risk may be too high
Vast resource for new ideas/content	Potential increase in: (a) contractual obligations, (b) insurance coverage issues, and (c) liability and damages exposure

(Marc Lieberstein, 2012)

consumer expectations. The crowd shaping also helps out in developing various new financial and insurance products based on the expenditure as well as saving habit of users. Presume that you are watching TV wearing the wrist badge which can sync the data about your feelings and the programs on the TV can be changed automatically as per your thoughts. This will be called as crowd shaped television.

The future retail stores, hospitals, multiplexes, colleges, concerts can be reshaped and redesigned with the help of crowd shaped experience. Crowd shaped digital advertising screens showing only those advertisements in which an individual customer or a crowd will prefer. The in-store music will be crowd shaped which will change automatically as per the preference of customers. An indigenously designed speaker 'Chune' from 'Clearleft' that creates crowd sourced playlists is a new social music concept designed for parties, office and other shared use settings. Another product is a jukebox 'Checkin DJ' which also arranges the music at any concert to match with the preferences and moods of crowd. Recently in Helsinki the department of transport has launched a crowd shaped mini - bus service named 'Kutsuplus' which provides services on demand bus services and can changed the routs the optimum basis.

The Philips and China Electronics Corporation (CEC) partnership, announced in July 2013, marked an attempt to develop and implement intelligent road and street lighting in cities across China. The joint venture will focus on LED fixtures and lighting management, including Philips CityTouch, a web - based system that allows streetlights to be used when and where they are needed. Scientists at the Chinese Academy of Science have unveiled a 'smart' window that can save and generate energy. Coated with temperature sensitive Vanadium Oxide (VO2), the window can regulate the amount of energy entering a building and store light energy in solar cells within the window frame.

The health sector will be the most benefitted with the crowd shaped products. Now, one can easily detect in advance that when and what type of disease he is going to suffer in next few days based on the movement of phagocytes in his body. 'Ktrack' the wrist watch which tells the level of glucose without any pain, 'Tricella Pill Box' reminds you about whether you have taken the prescribed pill or forgot. All these products are connected to your smart phone. Novartis has recently worked for crowd shaped Chronic Obstructive Pulmonary Disease (COPD) solution which may help more than 210 million people all around the world to tackle the breathing problem.

Several footballers have been died on the field due to various kinds of trauma. The Riddell Insite Football helmet contains sensors that alert coaches in event of trauma. In October 2013, the US soccer helmet manufacture 'Riddell' has developed an Insite Impact Response System. The sensor system embedded inside the helmet sends an alert to the doctor standing at side - lines in case of a substantial impact. The sudden heart - attack to drivers is very common problem causes severe accidents on roads. Ford has developed an intelligent car which keeps a tab on monitoring the driver's heart beat rate. In September 2015, Ford has unveiled the S - Max Concept, an intelligent multipurpose vehicle that includes multiple smart, care features. The driver's seat, which contains an ECG heart rate monitor, can help avoid mid - journey driver heart attacks and subsequent accidents. An onboard glucose level monitoring system alerts drivers of critical blood sugar levels. The vehicle is capable of communicating short - range via wifi with other similarly equipped cars, allowing warnings of road hazards to be relayed from car to car.

Canadian technology company OMsignal has announced plans to launch a compression shirt which can monitor an individual's heart rate, breathing and movement. Sensors woven into the fabric collect data which is then sent to the wearer's cell phone, where they can track and analyze information. Fully machine washable, the shirt is designed to be worn underneath garments or at the gym. Xkuty One is a smart electric scooter which automatically alerts the relatives of a rider in the event of a crash. Developed by Spain - based Electric Mobility Company, the bike - scooter hybrid includes an iPhone dock in the handlebars. Using the phone's gyroscope (which monitors orientation and momentum), Xkuty's app can detect crashes and automatically sends accident and location notifications to selected contacts. Xkuty One retails at around EUR 2800.

Created to help consumers reduce stress levels, PIP is a wireless biosensor developed in Ireland that reached its funding target on Kickstarter in July 2013. Users hold the device while playing a game (synced via Bluetooth to a monitor or smart phone), and it captures a Galvanic Skin Response from their fingertips in real - time. To succeed in a racing game for example, the player must be more relaxed than their competitor, as characters respond aversely to stress signals. The creators will release a Software Development Kit so others can develop apps. The Melon headband and companion app allows wearers to track, monitor and understand their mental focus during a range of activities. The headband measures brain activity using EEG and algorithms detect focus levels, using the data to give personalized feedback. Users can input external factors via the app to discover how time of day, weather conditions and environment all influence their ability to generate and sustain focus.

Another area where the crowd shaping is going to create some more milestones is the service sector led by food industry. ‘Emotinet’ a US company has developed an emotion recognition software with the help of USB cameras. This company provides services to P&G and some restaurants & food chains in form of real - time market research which led them to provide the customized food to the customers. Shadow is an app that allows users to record and remember their dreams. An escalating alarm clock wakes the user up gently, increasing the likelihood of dream recall, and the app prompts the user to record the dream via audio or text. Tracking dream and sleep patterns allows curious users to make connections between daily life and dream habits. Each entry is saved to a journal and can be anonymously pushed to the cloud, creating a global database of dreams, which shows users the sentiment and dream content of other participants around the world. Developed by Japanese product design brand Neurowear and unveiled in March 2013, the Mico headset consists of headphones and an EEG reader that rests on the wearer’s forehead and senses neural activity. Brain signals allow the device to detect the user’s mood (such as sleepy, stressed or focused), which is shown on an LED earpiece display. The headphones relay the information to a mobile running the Mico app and mood - appropriate music is selected from the database.

CONCLUSION

So much is there still to know, to discover and to find out the real potential of crowd shaping. Still no one is certain about real morphological features of crowd shaping yet is very promising. It will be another revolution in a conglomerate way where the technology again will be a dominant factor and used in all the spheres of each business sectors. It may have some limitations about breaching the privacy of users, more laws will have to be developed keeping in mind the ease of both parties of users as well as the of the companies. The life cycle of product will again shorten, the production will increase, service sector will boom and the employment will get a boost. The future lies in extraction of real time data making the more customized product and quicker delivery with ease of payment option. Be ready! You are being crowd shaped soon!

REFERENCES

Adobe. (2013). *Digital trends for 2013*. Retrieved 2016, from http://success.adobe.com/en/uk/programs/products/digitalmarketing/1301-28808-econsultancy-digital-trends-for-2013.html

Agarwal, N., Liu, H., Tang, L., & Yu, P. S. (2008, February). Identifying the influential bloggers in a community. In *Proceedings of the 2008 international conference on web search and data mining* (pp. 207 - 218). ACM. doi:10.1145/1341531.1341559

Blanc, B., & Figliozzi, M. (2016). Modeling the impacts of facility type, trip characteristics, and trip stressors on cyclists' comfort levels utilizing crowd sourced data. *Journal of the Transportation Research Board*, (2587), 100 - 108.

Chung, C. J., & Wee, H. M. (2011). Short life-cycle deteriorating product remanufacturing in a green supply chain inventory control system. *International Journal of Production Economics*, *129*(1), 195–203. doi:10.1016/j.ijpe.2010.09.033

Globescan. (2013). *2.5 Billion Aspirational Consumers Mark Shift in Sustainable Consumption*.

Halder, B. (2014). Evolution of crowd sourcing: Potential data protection, privacy and security concerns under the new media age. *Revista Democracia Digital e Governo Eletrônico*, *1*(10), 377–393.

Hoegg, J., Alba, J. W., & Dahl, D. W. (2010). The good, the bad, and the ugly: Influence of aesthetics on product feature judgments. *Journal of Consumer Psychology*, *20*(4), 419–430. doi:10.1016/j.jcps.2010.07.002

ISACA. (2013). *CGEIT Review Manual 2013*. Illinois: ISACA Publishing.

Kumar, V., Zhang, X., & Luo, A. (2014). Modeling customer opt-in and opt-out in a permission-based marketing context. *JMR, Journal of Marketing Research*, *51*(4), 403–419. doi:10.1509/jmr.13.0169

Lee Rainie, S. K. (2013). *Anonymity, Privacy, and Security Online*. Retrieved 2016 from http://www.pewinternet.org/2013/09/05/anonymity-privacy-and-security-online

Marc Lieberstein, A. T. (2012). *Crowd sourcing and Intellectual Property Issues*. Retrieved from http://www.acc.com/legalresources/quickcounsel/caipi.cfm

Mariani, A., Annunziata, A., Nacchia, F., & Vastola, A. (2016). Crowd sourcing in Wine Business: Co-Creation and Fundraising Experiences. In Successful Social Media and Ecommerce Strategies in the Wine Industry (pp. 93 - 114). Palgrave Macmillan US.

Meyerson, B. (2012). *IBM News Room*. Retrieved from http://www-03.ibm.com/press/us/en/pressrelease/39685.wss

Microsoft. (2013). *Internet of things - GfK*. Retrieved November 2013 from https://www.gfk.com/fileadmin/.../Internet_of_things_-_whitepaper_-_digitaal.pdf

Parkes, I. (2013). *Colemanparkes*. Retrieved 2016 from http://colemanparkes.com

Savitz, A. (2012). *The triple bottom line: How today's best-run companies are achieving economic, social and environmental success--and how you can too*. John Wiley & Sons.

Sorokin, A., & Forsyth, D. (2008, June). Utility data annotation with Amazon mechanical Turk. In *Proceedings of the IEEE Computer Society Conference on* Computer Vision and Pattern Recognition Workshops CVPRW '08. IEEE.

Stamford, C. (2013). Gartner Says Potential Size and Diversity of the Internet of Things Mask Immediate Opportunities. *Gartner*. Retrieved 2016 from http://www.gartner.com/newsroom/id/2564916

Sukaini A. K. M., & Zhang, J. (2016). Crowd sourcing and e-Commerce. Chinese Online Reviews.

Von Ahn, L., Ginosar, S., Kedia, M., & Blum, M. (2007). Improving image search with phetch. In Proceedings of the IEEE international conference on Acoustics, speech and signal processing ICASSP '07 (Vol. 4, pp. IV–1209). IEEE.

Zeithamal, V. A., Parasuraman, A., & Malhotra, A. (2002). Service quality delivery through web sites: A critical review of extant knowledge. *Journal of the Academy of Marketing Science*, *30*(4), 362–375. doi:10.1177/009207002236911

KEY TERMS AND DEFINITIONS

Big Data: An extremely large data set based on the consumption pattern, market trends, habits of consumers. Analysis of the big data helps to formulate various business policies.

Cloud Computing: Storing and retrieving of data from the internet based servers not from the local servers.

Crowd Shaping: The customized product based on the data gathered in form of crowd sourcing.

Crowd Sourcing: The data set collected from a large crowd (Paid or non-paid both).

Database: A collection of data from various sources about habits of consumers.

Internet of Things: A transfer of data without human involvement, through the network of machine, building and any live creature.

Real Time Data: The data set based on the current information.

Sensor: The technique to receive the signals.

Chapter 4
Reshaping Human Capital Formation Through Digitalization

Sulaiman Olusegun Atiku
University of KwaZulu-Natal, South Africa

ABSTRACT

This chapter focuses on reshaping intellectual capital formation via electronic-based learning platforms. A critical examination of the literature on human capital development through e-learning was conducted and it was found that digitalization enhances teaching / learning processes and activities, rather than rendering the traditional methods obsolete. The commonly used learning management systems are Blackboard, Class-Front and WebCT. With various virtual learning platforms such as game-based learning, mobile learning, social learning, and virtual world learning, the teaching and learning environments are being extended. The evolution of high levels of sophisticated information technologies across the globe has tremendously improved intellectual capital formation through digital collaboration, and interactions. Therefore, it takes continuous update of intellectual assets through digitized processes to keep abreast of vast innovations and technical know-how.

INTRODUCTION

The principal aim of this chapter is to present a report on intellectual capital development via electronic-based learning. The idea of reshaping intellectual capital formation through digitalization has become more prominent as a result of

DOI: 10.4018/978-1-5225-3191-3.ch004

technological innovations. The evolution of various electronic gadgets in educational industries across the globe renders traditional teaching and learning platforms obsolete. Organizational development (OD) practitioners, higher education institutions and individuals should be flexible in responding to the relevant complexities and technical innovations. Technological innovations render previously acquired (old) skills or competencies obsolete within a short period (Allen & Velden, 2012), suggesting the need for continuous update of skills and dynamic capabilities (Atiku, Chitakunye & Fields, 2014). In this digital age, skills and competencies can be easily upgraded through digitalization. The traditional approach to human capital formation could be upgraded by taking advantage of technological advancements in the 21st century.

Teaching and learning at higher institutions, as well as training and development interventions at corporate universities, are becoming more digitized to disseminate knowledge effectively in the global economy. For example, virtual reality of online learning environment are being explored using smart glasses as an effective way of enhancing human capital formation in the digitalized economy. In this context, distance is no longer a barrier for teaching and learning in the 21st century. Vital knowledge is speedily disseminated across all the continents, putting IT/internet or digitalization at the centre stage (Helbing, 2013, Pappas, 2017). One of the fundamental challenges facing developing countries is the cost of acquiring infrastructural facilities for knowledge creation through digitalization by individuals and institutions of higher learning. In specific terms, African countries are confronted with poor infrastructural facilities essential for e-learning. Unlimited internet connection as a basic means of livelihood in the Western countries remains a luxury in many African countries. In spite of the cost implications of an electronic-based teaching and learning platform, it remains one of the most effective ways of enhancing intellectual capital development across the globe (Salas & Gelfand, 2013; Sambamurthy, Bharadwaj & Grover, 2003).

This chapter seeks to uncover the benefits of intellectual capital formation through digitalization. In view of the huge capital requirement for full digitalization of teaching and learning in the 21st century, a blended learning approach, which is a combination of face-to-face or classroom teaching and online learning, might be more appropriate for developing countries in Africa. The author does not subscribe to the idea of total eradication of traditional teaching and learning methods in the 21st century. It is submitted that traditional teaching and learning approaches could be upgraded through digitalization to improve the competence-based levels required in this millennium. It takes continuous update of intellectual assets through digitized processes to keep abreast of vast innovations and technical know-how.

BACKGROUND

The concept "human capital" was introduced by an economist named Theodore Schultz in the 1960s to explain the present value of past investment in an individual's education, training and development (Investopedia, n.d.). The concept measures the economic value of individuals' skills or competencies acquired through long-term educational development. The human capital theory, which was widely promoted by Becker in the 1960s (Kucharčíková, 2011), submits that investment in employees'/ individuals' training and development, experiences and abilities have economic value for individuals, employers and the economy as a whole (Investopedia, n.d.). The economic value, which is the return on investment for employees, is their wages and salaries. Organizational productivity and profitability is the return on investment in training and development for an employer (Bhattacharya, Harold-Doty & Garavan, 2014; Percival, Cozzarin & Formaneck, 2013). At macro-level, economic growth and development is the return on investment in education for a nation (Hava & Erturgut, 2010; Olaniyan & Okemakinde, 2008). According to Oketch (2006), return on investment in education, learning and development of human resources can be grouped into two categories: financial and non-financial returns. Financial returns are measured in terms of the value of the investment and the monthly or annual earnings of human resources after such learning and development. On the other hand, non-financial returns relate to non-monetary returns on investment in education, learning and development of human resources, including community development services, political development and consciousness of civil rights.

The technological revolution that transformed the methods of human capital formation can be traced back to the mid-20th century (Forrestera, Ustinovaa, Kosyakovaa, Ronzhinab & Suraevac, 2016). The insight gained from the revolution was that access to information, experiences and technological innovations are valuable resources and major determinants of business (economic) performance (growth) in the global economy. The technological innovations in the 21st century have become so vast that digitalization has become the most suitable method to keep abreast of the required knowledge and competencies. In certain instances, the traditional methods of knowledge disseminating are not suitable in conveying the technical know-how required for the latest technological innovation (Chau et al., 2013). For example, gamification and virtual reality application is the most appropriate pedagogical technique used for pilots' training, which is usually sophisticated and very costly in terms of the software. Business simulation as an experiential learning approach is being advanced in recent times through gamification and virtual reality applications (Pappas, 2014; 2017). Therefore, there is a need to upgrade training content and techniques in line with the knowledge, skills, and abilities required for excellent service delivery in the 21st century. In addressing this gap, this chapter centres on reshaping human capital formation through digitalization. The next heading explains e-learning and electronic devices required for effective human capital development through e-learning.

E-LEARNING AND ESSENTIAL TECHNOLOGY

The term e-learning stands for electronic learning, meaning internet-based training or web-based training. According to the Learning Skills Development Agency (LSDA, 2015), e-learning is "the use of electronic technology to deliver, support and enhance teaching and learning". This definition shows that information technology plays a crucial role in providing, reinforcing and enhancing human capital development in both academia and industry. Similarly, e-learning is "the use of new multimedia technologies and the Internet to improve the quality of learning by facilitating access to resources and services as well as remote exchanges and collaboration" (Alonso, Lopez, Manrique & Viñes, 2005; eEurope, 2005). Evidently, e-learning is a form of digitized learning process developed to enhance the traditional form of knowledge sharing and transfer. This definition is considered an educational paradigm-oriented definition of e-learning (Sangrá, Vlachopoulos & Cabrera, 2012). Sangrá et al. (2012) provide a more acceptable definition of e-learning "as a form of learning which exploits the electronic media as a tool for improving the quality of teaching and training". The definitions above do not address the interactive aspect of e-learning (Woo & Reeves, 2007). Meaningful interaction between learners and instructors, or between a learner and other learners, is important in any learning process. According to Paulsen (2003), e-learning depends on the use of an interactive learning platform whereby learning contents are presented online and automatic feedback is provided regarding the activities of the learners.

Access to electronic devices such as computers as well as reliable internet connections is an essential facility for the e-learning process. In the case of mobile learning (m-learning), the learning process requires the utilization of mobile devices such as mobile phones, tablets, notebooks, handheld computers and other mobile technologies (LSDA, 2015; Naidoo, 2017). These devices are often used in combination with traditional training methods in education institutions and corporate universities. E-Learning and m-learning devices can be used to access e-books and other learning materials, and to interact with instructors as well as other learners. E-Learning content is usually conveyed through the internet, intranet or extranet, audio-or video-tape satellite TV and CD-ROM (Beal, n. d.; Nagarajan & Jiji, 2010; Naidoo, 2017; Shukla & Agrawal, 2013). E-learning activities can be conducted offline by completing assignments, listening to audio tape of lecture series, watching training videos, and rehearsing capabilities that were acquired at training sections (Chang, 2016). In these offline forms of e-learning, internet connections are usually not as important as electronic devices such as computers, tablets and other mobile technologies. The use of mobile technologies for online and offline learning activities gives learners the opportunity to participate in learning activities and digital interactions on group assignments and projects in their comfort zones.

The processes advanced for remodeling human capital development are web-based learning, computer-based learning, virtual learning and digital collaboration (Nagarajan & Jiji, 2010). Virtual learning and digital collaboration are made possible to learners via e-learning and m-learning. With virtual learning, a learner can actively engage in learning activities from anywhere (Chau et al., 2013). Virtual learning platforms such as Second Life and Active World are easily accessible to learners and used to promote distance learning by educational institutions. Also, experiential learning can be advanced through virtual world learning by simulating the real-world activities so that learners could learn in a simulated environment that is close to real events (Chau et al., 2013; Pallud, 2016). It offers opportunities to learners to learn real world activities without being exposed to real event repercussions (Chau et al., 2013). The world can indeed become a global village through virtual learning and digital collaboration, provided there are reliable internet connections and appropriate devices. The issue of distance and travelling expenses, costs for accommodation as limiting factor in the face-to-face classroom learning process is no longer a barrier to virtual learning and digital collaboration (Chang, 2016).

Digital learning management systems such as Blackboard, ClassFront and WebCT are commonly used platforms for enhancing teaching and learning by higher education institutions and corporate universities across the globe (Luck, Hashim & Din, 2012; Paulsen, 2003). For example, Blackboard can be used for submitting assignments to the course instructor using Assignment Board, while discussions and interactions among peers or between an instructor and learners are done using Communication Board. Digital interactions between a course instructor and learners are made accessible using a learning management system called Blackboard Collaborate. Another means of digital collaboration is a collaborative online innovative network (COIN). According to Gloor, Paasivaara, Schoder & Willems (2008, p. 1357), COINs are "cyber teams of self-motivated people with a collective vision, enabled by the Web to collaborate in achieving a common goal by sharing ideas, information, and work". A very good example of online innovative network where professionals or groups of highly motivated individuals share their thoughts, information and experiences is LinkedIn. The benefits of reshaping human capital formation through e-learning are presented below.

Benefits of Human Capital Formation Through E-Learning

Reshaping human capital formation by making good use of state-of-the-art technological innovations of the modern economy comes with many benefits to learners, educational institutions and industry. The obvious benefit of e-learning can be drawn from its definition (LSDA, 2015), which has to do with an enhancement of teaching and learning. The following are some of the advantages of e-learning extracted from scholarly articles:

- Digitalization of human capital development is cost-effective in the sense that it reduces travelling expenses and costs of hotel accommodation for learners on overseas training (Chang, 2016). The issue of distance is not a barrier in virtual learning.
- E-Learning enhances the efficacy of knowledge sharing, transfer and easy access to large amounts of data (Arkorful & Abaidoo, 2015). With e-learning large amounts of data can be compressed into electronic devices, which can be accessed in the learners' comfort zones.
- Web-based learning promotes learners' interaction with their instructors as well as peers (Paulsen, 2003). Some web-based learning is programd to provide automatic feedback to learners.
- Virtual learning by simulation offers opportunities to learners to learn real world activities without being exposed to real event repercussions (Chau et al., 2013). Corporate institutions adopt such programs for promoting experiential learning. Wastage of raw materials, defect products and loss of productive hours associated with on-the-job training can be minimized using computerized simulations.
- The adoption of e-learning allows self-pacing. For example, the asynchronous approach permits each learner to study at his or her own pace and speed (Arkorful & Abaidoo, 2015). The stress experienced by instructors in carrying both the slow and faster learners in traditional teaching and learning methods is eradicated using this approach.
- Learning activities can take place at any time and in any place. With mobile devices in particular, knowledge and information are more accessible to learners when and where it suits them.
- E-Learning offers access to professional counsel from experts across the globe through collaborative online innovative networks (Gloor et al., 2008). Learners can easily interact with the experts in their field of study from any part of the world (Devedžić, 2006). In spite of these benefits of reshaping human capital development through digitalization, it also comes with various disadvantages which are highlighted and explained below.

Issues in Human Capital Formation Through E-Learning

The advent of state-of-the-art technological innovations, Y2K compliance and globalization has transformed the business processes in various sectors worldwide, including the education sector. This renders traditional methods of teaching and learning obsolete in the 21st century. The developing and under-developed countries are expected to follow the pace of the developed countries in terms of electronic devices required for knowledge sharing and transfer. The fundamental disadvantage of e-learning is the cost of acquiring electronic gadgets for effective implementation of e-learning (Naidoo, 2017; Njenga & Fourie, 2010). At institutional level, budgetary restriction hinders the implementation of campus-wide e-learning solutions (Wagner, Hassanein & Head, 2008). Other disadvantages of e-learning are presented and explained below:

- Not all fields of study can adopt e-learning (Arkorful & Abaidoo, 2015). For instance, in applied sciences, it may be difficult to perform laboratory experiments using e-learning platforms. The use of e-learning is usually more appropriate in the fields of humanities, management and social sciences.
- Web-based learning may lead to network congestion and excessive usage of some websites which can also attract additional expenses on the part of the institution.
- Controlling or regulating bad activities such as cheating is a challenge, when tests for assessments in e-learning are done by proxy (Arkorful & Abaidoo, 2015).
- Degrees acquired through distance learning (virtual learning) are often underestimated compared to classroom education and/or blended learning (Devedžić, 2006).
- Adoption of e-learning requires additional costs for training the instructors and learners. Other expenses may include system maintenance (Barrett, 2011).
- Over-reliance on technology can cause some delays in delivery of teaching and learning. For example, technical problems in web-based learning can obstruct teaching and learning activities.
- Negative attitudes of instructors and learners toward the use of technology often hinder the implementation of e-learning (Devedžić, 2006). The reason is that internet-based communication usually takes more time than live communication in face-to-face teaching and learning. Therefore, stakeholders' awareness and support of e-learning are necessary prior implementation exercises.

INVESTMENTS AND STAKEHOLDER'S SUPPORT FOR E-LEARNING

One of the basic requirements for intellectual capital formation via digitalization is proper justification of investments in technologies required for web-based learning (McGill, Klobas & Renzi, 2014) or blended learning in some cases. Such investments should be adequately considered and approved by the management of educational institutions and/or organizational development practitioners. Other areas of consideration include the stakeholders' commitment to and support for implementation and maintenance of the program (Kisanga & Ireson, 2014; Wagner et al., 2008). Detailed explanations regarding institutional investment in technological infrastructure and other stakeholders' buy-in for the effective implementation of e-learning are provided below.

Investment in Technological Infrastructure

The investment in information and communication technology for proper implementation of e-learning is a fundamental requirement at institutional level. The areas of consideration may include the budgetary allocation for the procurement of electronic devices or technology for the enhancement of teaching and learning activities (Wagner et al., 2008). The budget for technology to implement e-learning will include system maintenance. Further analysis can be based on the rate of return on investment in electronic devices, development of pedagogical content and the primary benefit of virtual learning (McGill et al., 2014; Wagner, et al., 2008). Such approaches can be referred to as the cost-benefit analysis of investment in electronic-based teaching and learning. The return on investment could be in the short, medium, or long term. For a good-take up, the benefits should outweigh the costs. E-Learning initiatives also relate to vast technological innovations in which costs of regular system and software updates must be adequately taken into account (McGill et al., 2014). All relevant role players must be involved to avoid failure at the implementation stage.

The work of Kavanagh, Thite and Johnson (2015) shows that there are two major strategies for justifying investment in electronic-based learning: risk avoidance and institutional enhancement strategies. According to these authors, risk avoidance strategies are adopted when investments are perceived as mitigating against future risks that are likely to be faced by the institution; for example, the need to comply with the requirements of the accreditation bodies. The risk of losing accreditation may force an educational institution to seek funding to acquire necessary technology for the full implementation of e-learning. The teaching and learning enhancement strategies provide an estimate of the institution's expected levels of efficiency arising from the adoption of electronic-based learning (Kavanagh et al., 2015). The institutional enhancement strategies do not attract the risk of real loss if no action is taken apart from attempts to enhance the institution's competitive edge over its competitors in the industry. Besides the institutional enhancement strategy, the institutions need to ensure other stakeholders' awareness and buy-in for digital human capital formation.

Stakeholders' Awareness and Support

Sensitisation of all stakeholders is necessary, especially on the importance of electronic-based/blended learning in developing intellectual capital in the global economy (Kisanga & Ireson, 2014). Essentially this type of sensitization is intended to bring about stakeholders' awareness, commitment and support before take-up of the program (Wagner et al., 2008). Apart from approval and adequate investment by management, the commitment of the end users – such as course instructors and learners – is a determinant of effective implementation. To avoid resistance the level of awareness should extend to the provision of adequate training and /or workshops for the end users of e-learning. It is very important to seek other stakeholders' commitment and buy-in at the planning stage of the program.

According to Wagner et al. (2008), the enthusiasm and concerns of the stakeholders such as learners, instructors, educational institutions, content providers, technology providers, accredited bodies and employers are worthy of consideration. For instance, the willingness of instructional staff in creating more time to prepare and upload teaching and learning activities as well as material on virtual learning platforms is very important (Barrett, 2011). Parental support is also required in providing the necessary finances for learners to procure personal computers and electronic devices for easy accessibility of web-based teaching and learning activities and materials. Wagner et al. (2008) refer to learners as customers of e-learning. Based on global demand for high calibre of human assets, it is expected that learners will show more enthusiasms on the use of technology in acquiring essential skills for cutting edge innovations. Technology providers, as stakeholders of e-learning, need to ensure that technology standards are met. The various accreditation bodies are given the mandate to regulate the quality of teaching and learning activities that institutions offer their customers (Wagner et al., 2008). The customers in this context are different categories of learners in educational institutions. The accreditation bodies also assess the standards of the technology as well as levels of infrastructural facilities put in place by educational institutions, to advance intellectual capital formation in line with national or global requirements. Last but not least, the stakeholders are the employers of labor. Educational institutions should pay special attention to the requirements of the labor market in developing intellectual capital via e-learning platforms or a blended learning approach such that the quality of intellectual capital supplied by institutions can meet the demands of the labor market.

Dimensions and Delivery Methods of E-Learning

The advancement of educational technology has multiple effects on traditional teaching and learning methods in both developed and developing countries. The e-learning activities are characterized judging from its openness, flexibility and as a means of digital distribution of teaching and learning resources to different categories of learners across the globe (Khan, 2001).

The framework for e-learning initiatives could be seen as a guideline on implementation of e-learning by educational institutions and corporate universities established at various industries across the globe. The framework was designed on eight-dimensional action plans and sub-dimensional action lists.

The first dimension of the e-learning initiatives is pedagogical content, which involves considerations of setting the right learning objectives, teaching techniques, administration, strategies and platforms for the e-learning environment (Khan, 2001). Accordingly, electronic-based learning strategies include simulation, role play, tutorial, collaboration, demonstration and presentation, either audio or video. The second dimension of e-learning initiatives, according to Khan (2001), is the technological dimension. This dimension addresses issues relating to electronic gadgets in advancing teaching and learning in the digital age. Therefore, there should be adequate planning and procurement of computers and other required hardware and software. Thirdly, the interface design can be referred to as architectural structure of the e-learning platform, which may reflect the logo, colour, culture and motto of the institution (Khan, 2001; Petrou, 2011). These dimensions involve decisions relating to web-page design, content, navigation designs and test-running the website to ensure that it is user friendly.

The fourth dimension is based on concern regarding evaluation for e-learning, specifically in terms of learners' assessments, evaluation of learning instruction and environment (Ardito et al., 2006; Ozkan & Koseler, 2009). Management in this context is the fifth dimension for an electronic-based learning process, which involves maintenance of the teaching environment and knowledge distribution. The sixth dimension is resource support. This dimension encompasses online/offline assistance rendered to students, such as technical support, instructional support, counselling services and other online/offline support services to improve students' learning experiences (Khan, 2001).

Ethical considerations emerged as the seventh dimension in the framework for reshaping intellectual capital formation through web-based learning. The sub-dimensions of ethical considerations for web-based learning are focused on sensitive issues, related to learners' diversities such as socio-cultural diversity, information accessibility, etiquette and legal matters (Khan, 2001). The course instructors who lecture on distance learning platforms with an extensive global outreach need multi-cultural orientations and bigger global mind-sets as part of their leadership skills (Atiku & Fields, 2017). It becomes necessary for distance learning institutions to provide special training for instructors as a way of ensuring compliance with the ethical standards in the global learning environment.

The eighth dimension has to do with institutional issues classified into three sub-dimensions, namely administrative affairs, academic affairs and services to learners. The administrative issues worthy of consideration are managing organizational change, budgetary allocations to support the e-learning initiatives, return on investment, development and uploading instructions, marketing the programs and other administrative initiatives for efficient service delivery (Khan, 2001; Wright et al., 2014). Academic affairs refer to institutional action plans relating to the structure of the digital classroom and equipping the instructors with the necessary skills for lecture delivery on the platform (Almarabeh, Mohammad, Yousef & Majdalawi, 2014). Other academic matters are remuneration, consideration of intellectual property rights, instructional materials and class size (Khan, 2001). The sub-division services to learners, on the other hand, refers to institutional efforts in ensuring access to a digital library, e-books, digital interactions, campus Wi-Fi access to learning management systems, online tutorial services, orientation programs and other essential student services (Almarabeh et al., 2014; Wright et al., 2014).

Conversely, e-learning dimensions are dichotomized based on synchronicity, location, independence and mode (Khan & Badii, 2012; Ong, Lai & Wang, 2004; Wagner et al., 2008). Synchronicity in this context refers to learning attributes such as synchronous e-learning and asynchronous e-learning. According to the G-Cube (n. d.) classification these are the major types of e-learning. In asynchronous e-learning, teaching activities take place in real time (at the same time), whether instructors and learners are geographically dispersed or not (Wagner et al., 2008). With synchronous e-learning, virtual training activities take place at the same time while the facilitator and trainees are physically separated from one another. The delivery platforms in synchronous e-learning are listening to or watching a live radio/television broadcast, two-way live satellite broadcast, online lectures, video-conferencing and internet telephony. Asynchronous e-learning entails prepared instructions or programd tutorials, which are packaged and accessible to learners at their pace and time (Arkorful & Abaidoo, 2015; Wagner et al., 2008). The delivery platforms of asynchronous training are video-taped classes, self-paced training delivered through internet or CD-ROM, video web presentations or workshops, audio/video-taped lectures, automated feedback mentoring, computer-based training, e-mail and quick reference guide. Either synchronous or asynchronous learning dimension, an e-learning course encompasses instruction that is text-driven, interactive and simulated (Ferrimen, 2013). Examples of text-driven e-learning are PowerPoint slides and e-books involving texts, graphics and simple test questions. An interactive e-learning program is similar to a text-driven one in the sense that graphics, charts and diagrams are used for dialogue (Ferrimen, 2013).

The location dimension of e-learning is based on the consideration whether learners use an application at the same physical location as other learners and the facilitator or not (Khan & Badii, 2012; Wagner et al., 2008). The location dimension is illustrated using space, place and time. The independence dimension of web-based learning is open to two essential attributes, which are individual work and teamwork (Khan & Badii, 2012). The individualistic attribute is designed to ensure that each learner completes assigned tasks or activities independently. Teamwork or collaboration is designed to inculcate team spirit and principles of constructive interaction in learners through group projects and assignments that are executed electronically.

In deciding on the mode of e-learning institutions have two options:(1) pure, electronically-based teaching/learning activities and (2) a blended approach (Khan & Badii, 2012; Wagner et al., 2008). For example, pure e-learning platforms are usually used by institutions offering distance learning activities to learners across the globe. Almost all educational institutions that were involved with the traditional methods of intellectual capital development have reasoned along the blended learning approach. Therefore, conventional universities are currently adopting the blended learning approach. According to Singh (2003), blended learning is a combination of multiple delivery platforms that are designed to complement one another in enhancing teaching and learning activities. This approach also extends to a combination of face-to-face classroom teaching and electronic-based learning activities.

EMERGING E-LEARNING PLATFORMS

The multimedia used in enhancing learning and teaching process in recent times is communication technology, social networks and games (Lau, Yen, Li & Wah, 2014). Communication technology as a medium for e-learning enhances digital interactions among the stakeholders in the education industry. Social networks as a medium for e-learning extends the learning environments via s-learning (Lau et al., 2014) and m-learning (LSDA, 2015; Naidoo, 2017). The top six e-learning trends are gamification, automated course authorising, wearable tech training, big data, responsive learning management systems, and cloud-based systems (Pappas, 2016). The gamification of e-learning makes teaching and learning activities more interesting to learners, rather than the traditional abstract knowledge which can be boring. Game-based learning (GBL) as a paradigm shift in teaching and learning is advanced through a couple of technical innovations in "…multimedia interaction, computer graphics, human-computer interactions, and ubiquitous computing…" (Lau et al., 2014, p. 191). Accordingly, the main aim of GBL is to advance the teaching process through entertainment, and tasks that are visually and interactively oriented; targeted at achieving higher levels of understanding among learners.

The adoption of electronic-based learning content management systems (Simões, Rodrigues & De la Torre, 2013) is being propelled by vast innovations in information technology throughout the world. The evolution of numerous systems for integrating social networks has also paved the way for new generations of interactive learning systems in the knowledge economy (Lytras et al., 2015). The new generation of interactive learning systems is an extension of the collaborative learning system using social network systems (SNSs). The use of an SNS as an interactive learning platform is in line with the principles of social development theory proposed by Vygotsky (1978, cited in Adams, 2007). Vygotsky's theory proposed that social dialogue significantly influences intellectual development. This serves as a background of social network systems in enhancing intellectual capital development through digitalization. One can deduce that emerging knowledge and content management systems are platforms for enhancing teaching and learning activities, rather than abolishing the orthodox methods in the 21st century. With wearable gadgets, virtual reality (VR) online learning are enhanced through augmented reality (AR), which generates artificial objects in real environment for learners' interactions (Elton, 2017; Pappas, 2015; Timothy, 2016). The benefits of wearable tech training for organizations are immersive task simulation experience, detailed product knowledge transfer, efficient use of company resources, inconspicuous feedback, and better collaboration in workplace (Pappas, 2015).

Currently, digital collaboration is being synchronised using social media platforms such as Facebook group discussions, which are also accessible via mobile devices (Blewett & Hugo, 2016). Using social networks such as Facebook (Blewett & Hugo, 2016) and WhatsApp (Ashiyan & Salehi, 2016) by creating group chats per module will enhance learners' interactions and knowledge sharing capabilities on group assignments or projects via m-learning. The WhatsApp chats as a social network system, which has recently become very popular among learners, can also be used to extend mobile learning environments. The knowledge transfer and sharing capabilities of virtual communities of learners and instructors cannot be underestimated in the digital age.

New e-learning platforms are also emerging in radiology education for intellectual capital development in fields of studies where the use of images forms an integral part of teaching and learning (Xiberta & Boada, 2016). Accordingly, RadEd is a web-based framework with an excellent editor in creating image-based exercises for learners. The evolution of three-dimensional graphics in virtual world learning is effective in building real world learning experiences among learners (Chau et al., 2013). Through simulation, learners can reap the benefits of experiential learning activities without being exposed to any form of real world repercussions. Experiential learning via new generation virtual learning platforms substantiates the assumptions of constructivist learning theory that knowledge is being contextualised rather than acquired (Adams, 2007). According to this theory, learning is a process in which learners actively engage in creative activities based on an on-going knowledge and previously acquired experiences (Adams, 2007). Individual experiences influence the contextualisation of knowledge (Adams, 2007), which can be put to test via social network systems (Chau et al., 2013; Ellison & Boyd, 2013; Siemens, 2014). Owing to the evolution of high levels of sophisticated information technologies across the globe, digital collaborations and interactions have improved tremendously. Subsequently, a number of recommendations are put forward to individuals, OD practitioners, distance learning and conventional universities operating in the digital age.

SOLUTIONS AND RECOMMENDATIONS

There is no doubt that the development of sophisticated technological innovations has transformed business processes in many industries throughout the world. Consequently, there is need to reorganize the existing work structures, which begins with reshaping intellectual capital formation through electronic-based teaching and learning activities. This chapter proposes that continuous updating of intellectual assets through digitized processes is vital to keep abreast of the vast innovations and technical know-how in many industries across the globe. Therefore, OD practitioners; distance learning institutions and conventional universities world-wide need to take cognizance of emerging technologies aimed at enhancing virtual world teaching and learning activities. The reason is that digital age also requires the utilization of state-of-the-art technologies for intellectual capital development. For instance, emerging sophisticated virtual world learning platforms should be fully utilized for experiential training in various industries. This approach of staff development via simulation helps staff in attaining the right job experiences without being exposed to the real-world repercussion affecting the quality of products and services, as well as wastage of valuable raw materials (Chau et al., 2013; Pallud, 2016). Individuals, OD practitioners, management of services-and products-oriented organizations, as well as corporate universities world-wide would benefit from such an approach.

The higher institutions of learning should show greater levels of compliance with the labor market requirements by using the appropriate teaching and learning platforms in developing the right skills, competencies and dynamic capabilities. This would involve using technology in transforming abstract knowledge into experiential teaching/learning procedures in line with the expectations of employers of labor in the global economy. In practical terms, fine-tuning abstract knowledge into experiential teaching/learning activities can be intensified through a hi-tech blended learning approach (Chau et al., 2013). The students' creative and innovative capacity can be enhanced through digital collaboration and interaction, which can be promoted using m-learning, s-learning and other web-based learning platforms (Lau et al., 2014; LSDA, 2015; Naidoo, 2017). If adequately implemented, these platforms could be effective in extending the existing teaching and learning environments in the 21st century. Some research directions for the promotion of hi-tech teaching and learning platforms regarding further updates of intellectual capital in the global economy are proposed below.

FUTURE RESEARCH DIRECTIONS

It is believed that this chapter has extended the frontiers of knowledge regarding the possibilities of reshaping human capital formation through digitalization. The emerging e-learning platforms for enhancing teaching/learning processes were explored in this academic discourse, based on the labor market and knowledge economy requirements. Therefore, it is suggested that future research be sponsored on the development of artificial intelligence for human capacity building and employment generation by educationists and OD practitioners. Rather than over-emphasising the negative influence of artificial intelligence in various industries worldwide, efforts can be channeled or redirected towards extending human intelligence by means of artificial intelligence. Furthermore, critical thinking and applied research are essential to the process of discovering how artificial intelligence can be directed towards employment generation.

CONCLUSION

This chapter explores the benefits and challenges of reshaping human capital formation through digitalization. The evolution of state-of-the-art technological innovations has been instrumental in advancing teaching and learning environments. For example, social learning networks are growing through m-learning, s-learning and other learning management systems. It can be deduced that emerging learning and content management systems are platforms for the enhancement of teaching and learning activities and that the orthodox methods of teaching and learning should not merely be abolished in the 21st century.

Experiential learning is currently being demonstrated and advanced through the use of business simulation games and other virtual world learning platforms. Judging from the benefits of virtual world learning, it can be concluded that business simulation games, for example, expose learners to real world business situations without their encountering the actual business risks. Generally, virtual world learning assists learners in gaining the desired experiences without being exposed to the real-world repercussions. One of the fundamental challenges of intellectual capital development is that it requires high levels of investment in technological infrastructure for effective implementation. The electronic gadgets required for enhancing the teaching/learning process are "desktop, laptop, notebook, mobile and wireless tools, software, digital cameras, interactive whiteboards, e-communication tools (e-mail), virtual learning environments and other learning content management systems" (Naidoo, 2017, p. 181). Adequate funds should be budgeted for the acquisition of these kinds of technological infrastructure to reshape human capital formation through digitalization. Other challenges of digitalization of the teaching and learning process include getting full commitment and support of all stakeholders. Therefore, adequate planning and sensitization are required for stakeholders' buy-in and support of electronic-based learning.

REFERENCES

Adams, N. B. (2007). Toward a model for knowledge development in virtual environments: Strategies for student ownership. *International Journal for Social Sciences*, *2*(2), 71–77.

Allen, J., & Velden, R. K. W. (2012). *Skills for the 21st century: Implications for education. Meteor.* Maastricht University School of Business and Economics.

Almarabeh, T., Mohammad, H., Yousef, R., & Majdalawi, Y. K. (2014). The university of Jordan e-learning platform: State, students' acceptance and challenges. *Journal of Software Engineering and Applications*, *7*(12), 99–107. doi:10.4236/jsea.2014.712087

Alonso, F., López, G., Manrique, D., & Viñes, J. M. (2005). An instructional model for web-based e-learning education with a blended learning process approach. *British Journal of Educational Technology*, *36*(2), 217–235. doi:10.1111/j.1467-8535.2005.00454.x

Ardito, C., Costabile, M. F., De Marsico, M., Lanzilotti, R., Levialdi, S., Roselli, T., & Rossano, V. (2006). An approach to usability evaluation of e-learning applications. *Universal Access in the Information Society*, *4*(3), 270–283. doi:10.1007/s10209-005-0008-6

Arkorful, V., & Abaidoo, N. (2015). The role of e-learning, advantages and disadvantages of its adoption in higher education. *International Journal of Instructional Technology and Distance Learning*, *12*(1), 29–42.

Ashiyan, Z., & Salehi, H. (2016). Impact of WhatsApp on Learning and Retention of Collocation Knowledge among Iranian EFL Learners. *Advances in Language and Literary Studies*, *7*(5), 112–127.

Atiku, S. O., Chitakunye, P., & Fields, Z. (2014). Influence of organisational learning and human resource outcomes on commercial banks' performance in Nigeria. *J. Soc. Sci.*, *40*(1), 9–20.

Atiku, S. O., & Fields, Z. (2017). Multicultural Orientations for 21st Century Global Leadership. In N. Baporikar (Ed.), *Management Education for Global Leadership* (pp. 28–51). Hershey, USA: IGI Global. doi:10.4018/978-1-5225-1013-0.ch002

Barrett, B. (2011). Developing and implementing strategies to enhance intellectual capital in the online learning environment. In *Proceedings of the 3rd European Conference on Intellectual Capital ECIC '11* (p. 62). Academic Conferences Limited.

Beal, V. (n. d.). *E-learning-Webopedia Definition*. Retrieved January 12, 2017 from http://www.webopedia.com/TERM/E/e_learning.html

Bhattacharya, M., Harold-Doty, D., & Garavan, T. (2014). The Organizational Context and Performance Implications of Human Capital Investment Variability. *Human Resource Development Quarterly*, *25*(1), 87–113. doi:10.1002/hrdq.21182

Blewett, C., & Hugo, W. (2016). The learning affordance of a Facebook environment. In K. Pillay & C. Goodier (Eds.), *Teaching and Learning in the College of Law and Management Studies: Shared Approaches, Lesson and Good Practices* (pp. 46–63). Durban, South Africa: University of KwaZulu-Natal Publication.

Chang, V. (2016). Review and discussion: E-learning for academia and industry. *International Journal of Information Management*, *36*(3), 476–485. doi:10.1016/j.ijinfomgt.2015.12.007

Chau, M., Wong, A., Wang, M., Lai, S., Chan, K. W., Li, T. M., & Sung, W. K. (2013). Using 3D virtual environments to facilitate students in constructivist learning. *Decision Support Systems*, *56*, 115–121. doi:10.1016/j.dss.2013.05.009

Devedžić, V. (2006). Introduction to Web-Based Education. *Semantic Web and Education*, 1-28. eEurope (2005). *E-learning*. Retrieved January 12, 2017 from http://europa.eu.int/information_society/eeurope/2005/all_about/elearning/index_en.htmhttp://www.elearningconference.org

Ellison, N. B., & Boyd, D. (2013). Sociality through Social Network Sites. In W. H. Dutton (Ed.), *The Oxford Handbook of Internet Studies* (pp. 151–172). Oxford: Oxford University Press.

Ferrimen, J. (2013). *Three types of e-learning*. Retrieved January 18, 2017 from https://www.learndash.com/3-types-of-elearning

Forrestera, S. V., Ustinovaa, G. H., Kosyakovaa, I. V., Ronzhinab, N. V., & Suraevac, M. O. (2016). Human Capital in the Innovative Conditions. *Mathematics Education*, *11*(8), 3048–3065.

G-Cube. (n. d.). *Different types of e-learning*. Retrieved January 18, 2017 from http://www.gc-solutions.net/resources/articles/different-types-of-e-learning-and-what-suits-me-best.html

Gloor, P. A., Paasivaara, M., Schoder, D., & Willems, P. (2008). Finding collaborative innovation networks through correlating performance with social network structure. *International Journal of Production Research*, *46*(5), 1357–1371. doi:10.1080/00207540701224582

Hava, H. T., & Erturgut, R. (2010). An evaluation of education relations together with technology, employment and economic development components. *Procedia: Social and Behavioral Sciences*, *2*(2), 1771–1775. doi:10.1016/j.sbspro.2010.03.982

Helbing, D. (2013). Globally networked risks and how to respond. *Nature*, *497*(7447), 51–59. doi:10.1038/nature12047 PMID:23636396

Investopedia (n.d). *Human capital*. Retrieved December 10, 2016 from http://www.investopedia.com/terms/h/humancapital.asp

Kavanagh, M. J., Thite, M., & Johnson, R. D. (2015). Human resource information systems: Basics, applications and future directions (3rd ed.). New York: Sage Publishers.

Khan, B. H. (2001). A framework for web-based learning. *Web-based training*, 75-98.

Khan, K. U., & Badii, A. (2012). Impact of E-learning on higher education: Development of an e-learning framework. *Life Science Journal*, *9*(4), 4073–4082.

Kisanga, D. H., & Ireson, G. (2014). Challenges and strategies on adoption of e-learning in Tanzanian higher learning institutions: lessons to future adopters. In *Proceedings of the International Conference on Information Communication Technologies in Education ICICTE '14*, Kos, Greece (pp. 5-8).

Kucharčíková, A. (2011). Human Capital – definitions and approaches. *Human Resources Management & Ergonomics*, *5*(2), 60–70.

Lau, R. W., Yen, N. Y., Li, F., & Wah, B. (2014). Recent development in multimedia e-learning technologies. *World Wide Web (Bussum)*, *17*(2), 189–198. doi:10.1007/s11280-013-0206-8

LDSA (Learning Skills Development Agency). (2015). *New trends in education and learning-online learning*. Retrieved January 12, 2017 from http://www.lsda.org.uk/new-trends-in-education-and-learning-online-learning.html

Luck, L. T., Hashim, F., & Din, S. Z. M. (2012). A Creative and Literary Writing Digital LMS in Supporting Writers Group Learning and Knowledge Sharing among Creative Writers. *Procedia: Social and Behavioral Sciences*, *67*, 238–249. doi:10.1016/j.sbspro.2012.11.326

Lytras, M. D., Mathkour, H. I., Abdalla, H., Al-Halabi, W., Yanez-Marquez, C., & Siqueira, S. W. M. (2015). An emerging–Social and emerging computing enabled philosophical paradigm for collaborative learning systems: Toward high effective next generation learning systems for the knowledge society. *Computers in Human Behavior*, *51*, 557–561. doi:10.1016/j.chb.2015.06.004

McGill, T. J., Klobas, J. E., & Renzi, S. (2014). Critical success factors for the continuation of e-learning initiatives. *The Internet and Higher Education*, *22*, 24–36. doi:10.1016/j.iheduc.2014.04.001

Nagarajan, P., & Jiji, G. W. (2010). Online educational system (e-learning). *International Journal of u-and e-Service. Science and Technology*, *3*(4), 37–48.

Naidoo, V. (2017). E-Learning and Management Education at African Universities. In N. Baporikar (Ed.), *Management Education for Global Leadership* (pp. 181–201). Hershey, USA: IGI Global. doi:10.4018/978-1-5225-1013-0.ch009

Njenga, J. K., & Fourie, L. C. H. (2010). The myths about e-learning in higher education. *British Journal of Educational Technology*, *41*(2), 199–212. doi:10.1111/j.1467-8535.2008.00910.x

Oketch, M. O. (2006). Determinants of human capital formation and economic growth of African countries. *Economics of Education Review*, *25*(5), 554–564. doi:10.1016/j.econedurev.2005.07.003

Olaniyan, D. A., & Okemakinde, T. (2008). Human capital theory: Implications for human resource development. *Pakistan Journal of Social Sciences*, *5*(5), 479–483.

Ong, C. S., Lai, J. Y., & Wang, Y. S. (2004). Factors affecting engineers' acceptance of asynchronous e-learning systems in high-tech companies. *Information & Management*, *41*(6), 795–804. doi:10.1016/j.im.2003.08.012

Ozkan, S., & Koseler, R. (2009). Multi-dimensional student's evaluation of e-learning systems in the higher education context: *An empirical investigation. Computers & Education*, *53*(4), 1285–1296. doi:10.1016/j.compedu.2009.06.011

Pallud, J. (2016). Impact of interactive technologies on stimulating learning experiences in a museum. *Information & Management*.

Paulsen, M. F. (2003). Experiences with Learning Management Systems in 113 European Institutions. *Journal of Educational Technology & Society*, *6*(4), 134–148.

Percival, J. C., Cozzarin, B. P., & Formaneck, S. D. (2013). Return on investment for workplace training: The Canadian experience. *International Journal of Training and Development*, *17*(1), 20–32. doi:10.1111/ijtd.12002

Petrou, A. (2011). Assessing Integration of Infrastructure 2.0 Tools in Open Learning Websites: Highlighting a Need for Purposeful Design. In Proceedings of Informing Science & IT Education Conference (pp. 427-441).

Salas, E., & Gelfand, M. J. (2013). Introduction to the Special Issue: Collaboration in multicultural environments. *Journal of Organizational Behavior*, *34*(6), 735–738. doi:10.1002/job.1880

Sambamurthy, V., Bharadwaj, A., & Grover, V. (2003). Shaping agility through digital options: Reconceptualizing the role of information technology in contemporary firms. *Management Information Systems Quarterly*, 27(2), 237–263.

Sangrà, A., Vlachopoulos, D., & Cabrera, N. (2012). Building an inclusive definition of e-learning: An approach to the conceptual framework. *The International Review of Research in Open and Distributed Learning*, *13*(2), 145–159. doi:10.19173/irrodl.v13i2.1161

Shukla, V. K., & Agrawal, P. (2013). Towards Semantical E-Learning from E-learning. *International Journal of Science and Research*, *2*(11), 243–247.

Siemens, G. (2014). *Connectivism: A learning theory for the digital age*. Retrieved January 24, 2017 from http://er.dut.ac.za/handle/123456789/69

Simões, T., Rodrigues, J. J., & De la Torre, I. (2013). Personal Learning Environment Box (PLEBOX): A new approach to E-learning platforms. *Computer Applications in Engineering Education*, *21*(S1), 100–109. doi:10.1002/cae.20537

Singh, H. (2003). Building effective blended learning programs. *Educational Technology-Saddle Brook Then Englewood Cliffs NJ*, *43*(6), 51–54.

Wagner, N. L., Hassanein, K., & Head, M. M. (2008). Who is Responsible for E-Learning Success in Higher Education? A Stakeholders' Analysis. *Journal of Educational Technology & Society*, *11*(3), 26–36.

Woo, Y., & Reeves, T. C. (2007). Meaningful interaction in web-based learning: A social constructivist interpretation. *The Internet and Higher Education*, *10*(1), 15–25. doi:10.1016/j.iheduc.2006.10.005

Wright, A. W., Hamilton, B., Mighty, J., Scharfe, E., Muirhead, B., & Vail, S. (2014). Shared Modular Course Development: A Feasibility Study. *Centre for Teaching and Learning Reports*. Retrieved January 22, 2017, from http://scholar.uwindsor.ca/ctlreports/1

Xiberta, P., & Boada, I. (2016). A new e-learning platform for radiology education (RadEd). *Computer Methods and Programs in Biomedicine*, *126*, 63–75. doi:10.1016/j.cmpb.2015.12.022 PMID:26774237

KEY TERMS AND DEFINITIONS

E-Learning: Electronic-based learning involves the use of electronic gadgets in delivery and in supporting and enhancing teaching and learning activities.

Game-Based Learning: In this approach to learning, interesting and game-like human-computer interactions are used in enhancing experiential learning in the digital age.

Learning Management Systems: These are online learning platforms used in advancing teaching and learning activities as well as internet-based interaction.

M-Learning: Mobile learning is the adoption of mobile devices such as mobile phones, notebooks, tablets and other handheld computers in enhancing digital interactions by extending teaching and learning environments.

S-Learning: Social learning extends learning environments and enhances collaboration with the use of social network systems.

Virtual World Learning: The computerized software engaged in advancing experiential learning by simulating real world activities so that learners could learn in a simulated environment close to real events.

Web-Based Learning: This is an internet-based teaching and learning platform mostly used by distance learning institutions to integrate geographically dispersed learners across the globe.

Section 2

Impacts of Digitalization on Work Structures

Chapter 5
Organizational Intervention Plan for Automation SMEs:
Case Study Gisitca

José G. Vargas-Hernández
University of Guadalajara, Mexico

Ángel Daniel Rodríguez Ortega
Universidad Cristóbal Colon – Calasanz, Mexico

ABSTRACT

This study has the aim of identify the main causes of a bad work environment with a high rate of turnover. The objective is to propose an intervention plan to increase the participation, commitment and employees' proactivity. This job is performed with a case study with the quantitative paradigm, transversal and exploratory; the selected sample is from a PYME dedicated to automation power services. For it is based on the model of situational leadership Hersey and Blanchard, in addition to job satisfaction survey NTP 213.

INTRODUCTION

In recent years, several associations and public institutions in Mexico have been concerned about supporting the creation of small and medium enterprises (SMEs) which are the main source of employment in the country with 72%. It has been worrying the failure rate of these companies in their early years and alarming their life expectancy, as indicated by the National Institute of Statistics and Geography (INEGI, 2015) who mentions that 6 (0.64) in 10 service companies survive the first

DOI: 10.4018/978-1-5225-3191-3.ch005

year and has a life expectancy at birth of 8 years. The probability decreases as they are older enterprises and life expectancy tends to increase with increasing in size. Such companies have a mature age of 20 years and a range of employees from 31 to 100; only remain alive 65 of 100.

CONDUSEF (2013) states that the most common mistakes made by SMEs and that can be very costly to them is the lack of a strategic analysis. Mismanagement and the dreaded incompetence of the administrator, who usually is the owner, and who fails to take over the business because often undertakes the business and makes decisions according to his gut feeling, also is not prepared to have his staff. That is, no lead in his own company.

This study aims to propose a plan of action to increase participation, commitment and proactivity of employees of family SMEs Business Process Automation (BPA) service provider, GISITCA case study.

This research consists of six sections. The first describes the background and origin of the problem. The second is delimited and defined the problem. In the third the realization of this project is justified. In the fourth, the methodology used to support research is established. In the fifth diagnostic results are analyzed and the intervention plan is designed. And finally, in the sixth section, the conclusions and recommendations arise.

Finally, I want to thank the support and facilities provided by the company GISITCA and to one of the authors for his valuable recommendations for the completion of this project.

BACKGROUND OF THE PROBLEM

Family businesses are the most common organization type in the world. In fact, most large companies were born as a family business (Lee, 2004). In Mexico micro, small and medium enterprises (SMEs) are the backbone of the country's economy since they are the main source of job creation, that is according to data provided by the National Institute of Statistics and Informatics (INEGI): of the 4 million 15 thousand companies, 99.8% are SMEs which corresponds to 52% of GDP and 72% of jobs in the country (PROMEXICO, n. d.) Belausteguigoitia (2003) mentions that in Latin America 9 out of 10 companies are family, which dominate most industries and are led by their founders (e.g. Grupo Carso, FEMSA, Televisa, etc.).

A determining factor in the growth of SMEs is the role of leader, who must direct the course of the company and to overcome the difficulties that may arise. Usually, the founder of the company owner is his leader exerting a paternalistic leadership style. He decides everything about the business and is responsible for the success or defeat of this. He must overcome luck in a sea of uncertainty of a globalized market, where technology has changed the way production and marketing of products. Given this scenario companies and especially SMEs require leadership that is an agent of change (Belausteguigoitia, 2003).

SME Integrator Group of Integrator Solutions and Technological innovations Control systems, enhancement and BPA (GISITCA) is a Mexican company founded in 1999. It specializes in the design, development and manufacture of control systems and power while systems integration PLC - based BPA is oriented primarily to serve the electricity sector of the domestic industry. Currently its workforce is made up of 45 people.

Since its inception, GISITCA has been characterized as a leader in the research and development of new and better technologies to help promote and ensure excellent quality of products and services; for the proper functioning of industries, through studies, analysis and engineering design; implementing appropriate and necessary equipment that requires the client to a high degree of reliability in the management of its electricity. In 2011 GISITCA and Casas Co. created GC Power to offer its services to the US industry based in Houston, TX.

Currently the company is in the process of expansion. However, since late 2014 it has been detected deterioration in the working environment, causing an increase in turnover, which is an alarm signal for the direction. Staff middle managers with extensive experience they have left the company, being affected by delays in the completion of projects. In turn, there is an overload of work while is integrated new personnel which causes discomfort and tension in the team.

DEFINING THE PROBLEM

GISITCA is currently facing a media turnover and increased operational controls that can be detrimental to the economy of the company. Also, the company is about to face a potential growth due to the approval of the energy reform. However, it must be prepared efficiently Thus, the following research question arises - How to improve the participation, commitment and proactivity of employees?

JUSTIFICATION

Until the energy reform of 2013, the contribution of the private sector in the electricity sector was only allowed in the form of Independent Power Producer (IPP) and self - sufficiency. This means that after covering the needs of the generator, it can sell the surplus energy only to the Federal Commission of Electricity (FCE). Only in 2011 the IPPs accounted for 29.1% of gross electricity generation (CFE, 2012).

With energy reform is expected there is an increase in energy infrastructure through private initiative. The above will result in increased demand for BPA in this sector. In turn, participation of foreign companies is expected imminent in supply service, representing that GISITCA compete with them.

With a proper intervention, it is expected the leadership design styles of leadership commensurate direction according to the challenges that it faces, promoting a better work environment, reducing turnover by having employees committed to the goals of the company and leaning on a system of labor motivation So that staff have greater participation and initiative.

Conceptual Framework

At present, organizations try to survive in a highly competitive environment, characterized by a market with discontinuous and unpredictable changes, which demands flexible companies and therefore of light structure in which its members have greater responsibility and commitment to the objectives of the organization. In this way, management should seek to develop the maximum performance of staff so that they can efficiently carry out their activities in an attractive work environment where they have rewarding experiences (Delgado & Delgado, 2003).

With a proper intervention, it is expected the management design styles of leadership and commensurate direction according to the challenges that it faces, promoting a better work environment, reducing turnover by having employees committed to the goals of the company and leaning on a system of labor motivation so that staff have greater participation and initiative.

Concepts

1. **Work:** Before starting to talk about the subject of motivation and job leadership, it is essential to briefly describe the concept of work, to facilitate the subsequent understanding of these issues. The Real Spanish Academy (Real Academia Española, 2017) defines work as the human effort applied to the production of wealth, as opposed to capital.

In this document, it is adopted the ecliptic concept of work as the effort (physical or mental) of man in developing transformation activities, to obtain remuneration to meet human needs. An important aspect of a job well done is that it indicates if the staff and their choice for the position are appropriate. These efficiency criteria are used to grant promotions, prizes, and incentives, in addition to identifying inefficient staff to train them better or take action on them.

2. **Motivation:** There are several approaches to try to define the concept of motivation. Historically different disciplines have contributed information to the subject, such as psychoanalytic theory, philosophy, social psychology, cognitive orientation, etc., having in ancient Greek philosophy its origin. Maslow (1943) stated that needs have a hierarchy of importance to be solved and these are ordered in a pyramid of five basic needs: basic or physiological, security, social, esteem and self - realization. In this order motivation is generated to cover the needs and the following is the imperative to satisfy. Hunt (1993) states that motivation is when a person tries to reach a goal, and to achieve it behaves in a certain way.

Robbins (2004) describes the concept of motivation as processes that account for the intensity, direction, and persistence of an individual's effort to achieve a goal. This definition is the one adopted in this paper, Intensity, direction and persistence, where intensity refers to the effort made by the person to achieve his goal, management refers to the effort being turned in the way of trying to reach the goal, and persistence indicates the firmness with which the person will maintain the same intensity until reaching the target.

3. **Work Motivation:** Motivation in the workplace refers to the process that triggers, guides and maintains the behavior of workers towards the completion of the expected objectives. This makes it important to identify the causes that stimulate human action, because managers must manage these elements so that their staff is satisfied and the organization performs adequately (López, 2005).
4. **Compensation:** In the organizational area, compensation refers to a reciprocal exchange of resources where both parties evaluate what they contribute and what they receive in return.
5. **System of Labor Rewards:** Wages and incentives support the implementation of strategies, which try to form the behavior of staff and the group. Reward systems should be consistent with organizational objectives and motivate employees to collaborate and cooperate in achieving those goals (García, Posada & Hernández, 2012). For Chiavenato (2002) the rewards system consists of a package of benefits and the mechanisms and procedures to grant them. The incentives can be: salaries, bonuses, vacations, promotions, positions of growth, among others.
6. **Organizational Climate:** There are a variety of studies on the work climate that have been developed since the 60s, which is why there are several definitions of this concept. Some of the definitions cited by Fumham (2001) are listed below -

a. Forehand and Von Gilmer (1964) define the concept of work climate as the set of characteristics that describe an organization and that distinguish it from other organizations. These characteristics are relatively enduring over time and influence the behavior of people in the organization.
b. Tagiuri (1968) defines organizational climate as a relatively enduring quality of the internal environment of an organization that its members experience and influence its behavior, and can be described in terms of the values of a specific set of characteristics or attributes of the organization.
c. For the present study, the definition proposed by Weinert (1985) is taken as the description of the set of stimuli that an individual perceives in the organization, which configure their working context.

7. **Leadership:** In the business sector, there is a generally accepted agreement that indicates that the success or failure of a company depends, to a large extent, on the quality of its leaders (Peiró, 2000). Several authors have given their point of view on leadership, among them are considered for this document the following:

For Tannenbaum, Weschler & Masarik, (1964) leadership is the interpersonal influence exercised in a situation and directed through the communication process towards achieving a target or targets (quoted in Yukl, 1994, p. 2). Hersey, Blanchard and Johnson (1998, p. 99) argue that leadership is the process of influencing the activities of an individual or group in efforts to reach a goal in a certain situation (quoted in Sánchez, 2010, p. 26). Jacobs and Jaques (1990) define leadership as a process of providing a purpose to the collective effort, and cause based on the willingness to achieve this effort (quoted in Yukl, 1994, p. 3).

This document describes the ecliptic concept of leadership which is used as the process of influencing an individual or group who, using their skills and expertise strive to meet the objectives of the organization.

Review of Theories

1. **Situational Leadership:** Paul Hersey and Ken Blanchard developed the theory of situational leadership in 1969. It is an important intuitive model in the situational approach to leadership effectiveness. It is the best-known model and used by managers in the US industry, because it can be applied in any situation where a person influences another leader (Sanchez, 2010).

The basis of situational leadership is the relationship between the type of leadership that a leader exerts and the degree of support that gives to the contributor. Likewise, it is the level of development that the employee exhibits in an activity and individual or group specific target. This provision of the employed may vary depending on the assignment. For example, a seller may have a high level of development for sales calls, but not to write proposals for clients. This means that the leader provides a great direction and supervision in developing proposals and little guidance and help in the calls (Benavides, 2004).

Little or excessive oversight negatively affects the development of individuals, it is necessary that the manager adapts leadership style (figure 1) according to the development which has contributor, being the essence of situational leadership I (raised in the 60s). In situational leadership II (updated in 2007), it assumes that individuals can and want to develop, but there is a leadership style that encourages development. Leadership must fit the situation (Blanchard, 2007). Situational leadership defines four quadrants that establish the kind of maturity employee and leadership style to be applied: Direction (S1), training (S2), support (S3) and delegation (S4).

Development levels are classified as enthusiastic beginner (D1), disillusioned learner (D2), capable but cautious performer (D3) and self - reliant achiever (D4). Their characteristics are shown in Figure 2 (Blanchard, 2007).

Figure 1. Leadership styles
Source: Blanchard (2013)

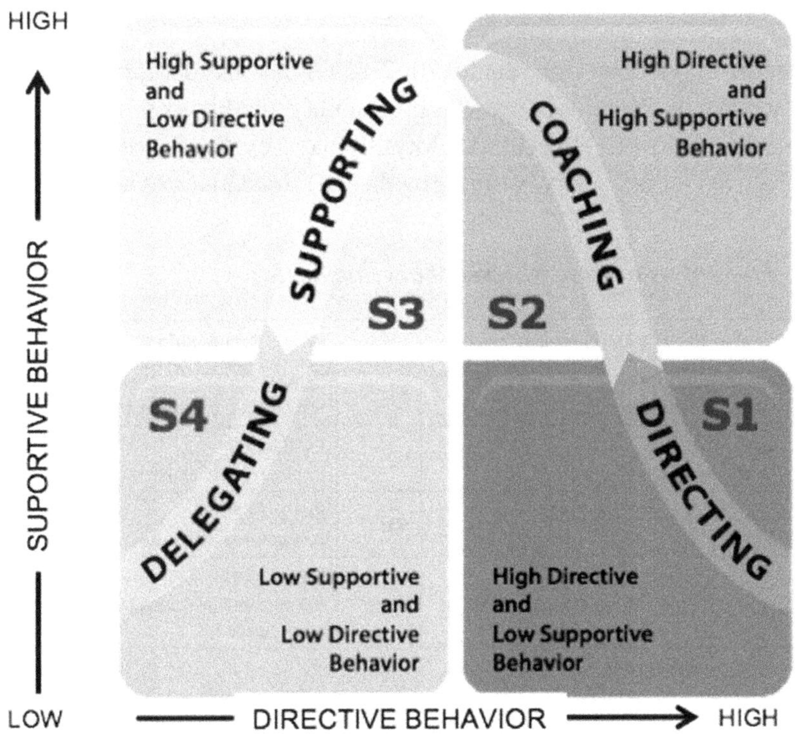

Figure 2. Levels of development
Source: Blanchard (2013)

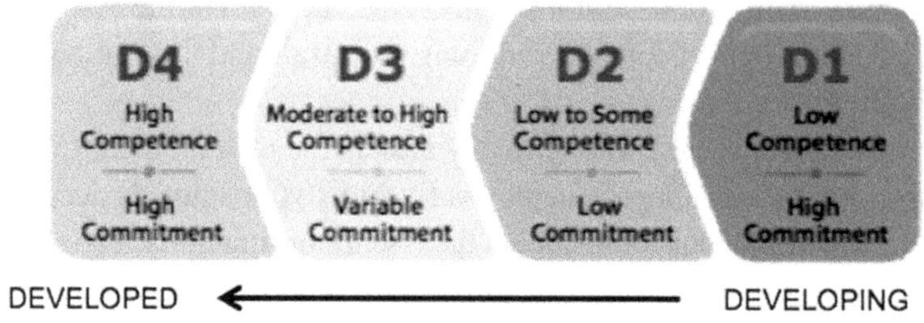

The meaning of the curve, from the perspective of the leader, is that according to the type of development or disposition of the employee (or group) to a target, a type of leadership will be applied as appropriate, as shown in Table 1.

The level of competence of an employee increases when strikes a target. The leader should begin his task of instructing or supporting until maturity of staff and their level is from moderate to above average. At this point the partner is motivated and committed, so the support of the supervisor requires not be the same and this will decrease, increasing the responsibility on the individual, this being a positive indicator of confidence (Benavides, 2004).

2. **Incentive Systems:** Chiavenato (2002) states that in order for the organization to achieve its objectives (growth, profitability, quality of production, etc.); it must know how to direct staff efforts so that they also reach their individual objectives (better salary, work, growth, etc.). Incentive payments are usually

Table 1. Types of leadership by level of development

Level of Partner Development		Leadership to Use	
D1:	Little competition Much commitment	**S1:**	LEAD Structuring, control and monitor
D2:	Some competition Less commitment	**S2:**	INSTRUCT Explain, guide and ask for suggestions
D3	Competition average to high Variable commitment	**S3:**	SUPPORT Recognition, listening
D4	High competition High commitment	**S4:**	DELEGATE Responsible decisions but giving support when needed

Source: Adaptation from Blanchard (2013)

awarded for specific performance results rather than seniority. Incentives are not the answer to all staff changes, and can do much more than increase worker performance.

Stoner, Freeman & Gilbert (1996) cited by Garcia, Posada & Hernández (2012) affirm that the organization when establishing an incentive system can be questioned: Are the bonds must be monetary or in actions? How are the results of the workers measured? What criterion will the managers apply to allocate the bonds? What amount is the bond? The answer is to align the incentive program with the objectives of the organization, keeping in mind that the staff adopts a behavior that produces rewards.

When creating a reward system, it should be open to feedback that supports positive behavior and thereby increase positive staff performance. For employees, the incentive is tangible or intangible gratification through which people are integrated into the organization (Chiavenato, 2002). The stimuli are offered to reinforce activities in the organization as: increase the responsibility of the person in the company, increase the interdependence between the individuals and the organization and finally incentivize the actions that give value to the organization (Chiavenato, 2002). For Garcia, Posada & Hernández (2012) performance and labor efficiency is influenced by four types of stimuli that affect the level of job satisfaction of the employee (figure 3):

- **Informal Rewards:** Nelson (1996) mentions that informal or spontaneous rewards are the initiative of management and are awarded on the basis of performance. Their objective is to reward and recognize the efficient work of employees. This stimulus to be successful, it is suggested to adjust the reward to the person, considering his personal tastes so that for him it is a truly satisfactory reward. Likewise, the prize must be adapted to what has been achieved, so that the reinforcement corresponds to the amount of money and effort used (time for planning and execution). Something important is that this type of stimulus must be granted in a timely manner to achieve its purpose.
- **Rewards for Specific Actions and Achievements:** Hersey and Blanchard (1997) define them as the rewards that some companies give to actions or ideas that are important to the whole organization, for example: ideas for economizing, providing an exceptional customer service, reaching some sales goal, etc. The most common recognition is the employee of the month. It is recommended that the selection of candidates should involve the staff and not only the managers (García, Posada & Hernández, 2012).

Figure 3. Diagram of rewards
Source: García, Posada & Hernández (2012)

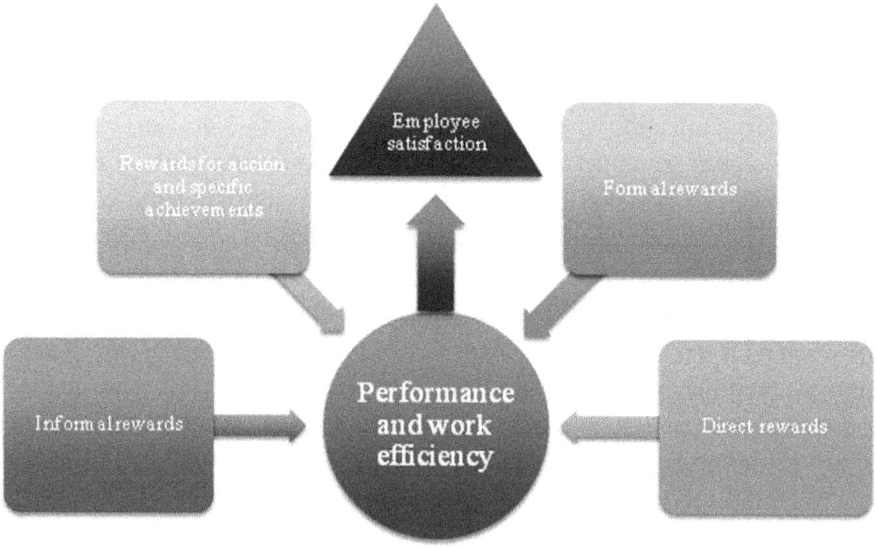

- **Formal Rewards:** They are the formal reward programs that the company develops under its initiative in order to keep its staff motivated, although some studies prove otherwise. It is suggested that the program be flexible and the rewards adapted to the needs of each employee. It is important that the rewards are timely and equitable to denote that they are fair, as well as enhancing the value of the awards when delivering them at an event (Nelson, 1996).
- **Nelson (1996) Suggests a Simple Rule to Maintain Staff Motivation:** for every four informal rewards obtained, a slightly more formal recognition should be given, and in turn, for every four formal rewards, something more formal and public should be given, and so on, until granting salary increases and promotions. It should be avoided to assign generalized incentives that can only discourage and impair the performance of the best employees because they do not feel especially recognized.
- **Indirect Compensation:** It refers to the economic incentives and services provided outside the benefits of the law, are social and personal development of employees, such as: life insurance, medical expenses insurance, birth of a child, food services from the company, support for academic preparation, nursery for children, among other benefits and additional services, which allows to meet some objectives of the company as a reduction of turnover rate, reduction of personnel conflicts, advantages for recruitment, etc. It is

also necessary to have a system of sanctions that is clear and includes the necessary disciplinary measures to guide the behavior of workers, so as not to deviate from the expected standards, to avoid repetition of behavior (verbal or written warning) and in the extreme case to punish recidivism (suspensions and discounts) and in the worst-case dismissal of the organization (Chiavenato, 2002).

3. **Evaluation of Results or Productivity:** The objective of evaluating the performance of the staff is to establish a standard or rule that allows its subsequent control, and thus to identify if the execution of the work is satisfactory, efficient or productive, so that the necessary corrections, such as training, change of activities, etc., or in the best of cases to recognize the best elements granting them ascents, prizes, incentives, etc. This measurement can also be useful to evaluate the efficiency of a department, area or company, as a basis for the payment of salaries per piece, calculate costs, etc. (Arias, Basualdo & Heredia, 1989).

Arias et al. (1989) establish as requirements of the evaluation measures the following:

- **Objectivity:** In setting the criteria should be objective and impartial.
- **Validity:** The criteria must show the effectiveness or inefficiency of the person judged as real as possible.
- **Reliability:** The criteria must be consistent, obtaining the best result every time it is applied on similar conditions of work and personnel.

4. **Balance Scorecard:** According to Kaplan & Norton (2005), traditional accounting indicators such as return on investment and earnings per share can be misleading for continuous improvement or innovation activities because the skills and competencies are outdated that the current business environment demands. Researchers have determined that measurement systems based on financial and operational indicators working in isolation are insufficient to measure productivity; this is because managers require a balanced presentation of both financial and operational measures (Kaplan & Norton, 2005).

The solution is the Balance Scorecard, a set of measurements that provides a fast and comprehensive view of the organization. It includes financial indicators that show results of actions already implemented, and is complemented by operational indicators with information on customer satisfaction, internal processes and innovation and organizational improvement activities that drive future financial performance (Kaplan & Norton, 2005).

The balance scorecard, as seen in Figure 4, shows information from four perspectives: financial, internal to the company, innovation and learning and customer perspective.

- From a customer perspective, the Balance Scorecard allows managers to translate the organization's mission into indicators and goals that reflect what is important to customers: quality, time, performance, service and cost. For this, it is necessary to rely on client evaluations (Kaplan & Norton, 2005).
- For the internal perspective of the company, customer - based measurements should become indicators of the crucial operations that the company must perform internally to meet customer expectations: quality time, productivity and cost (Kaplan & Norton, 2005).
- From the perspective of innovation and learning, the organization is in a global competition market, which forces it to make continuous improvements on existing products and processes and to launch new products. Innovation indicators focus on the ability to develop and introduce standard products that are the basis of future sales. Measuring manufacturing improvements focuses on new products rather than on improving the manufacturing of existing products (Kaplan & Norton, 2005).

Figure 4. Balance scorecard links performance measures
Source: Kaplan & Norton (2005)

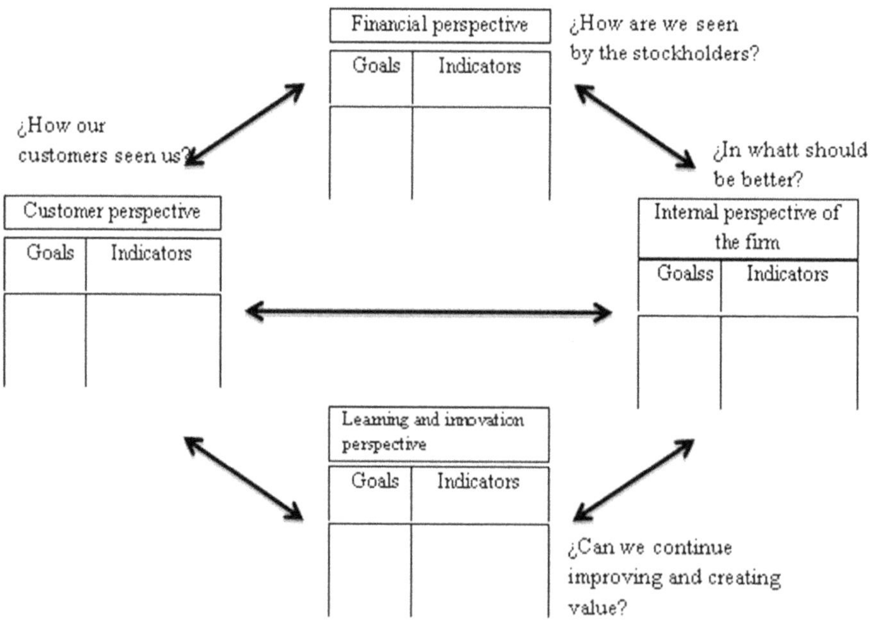

- For the financial perspective, the indicators show whether the company's strategy, its implementation and execution are contributing to improve the results that are important for the financial health of the organization: profitability, growth and shareholder value (Kaplan & Norton, 2005).

Empirical Review: Who Has Done Similar Research?

Cortes (1999) argues that among all major leadership approaches such as trait theory, behavioral, contingency, it stands the theory of situational leadership proposed by Hersey and Blanchard, which is the most widely used in business. In the academic field is challenged by scarce empirical evidence to validate their hypotheses and not consider important situational factors beyond the maturity of employees (Northouse, 1997; Yukl, 1989). This, combined with the instruments that are copyrighted, may be a reason why there are little empirical investigations of cases (even more in Mexico) that make use of situational theory II of Blanchard and the existing use adaptations of instruments. It is rarer still to find studies on BPA electricity companies.

In Spain, it can be cited the study entitled Determinants in the management of human resources in service companies that incorporate systematically new technologies: a case study in the Valencian community, dealing with changes in companies using the technology and human resources as set policies that contribute to improvements in using technology service companies. HR must ensure loyalty and belonging of employees and monitor employee performance. Among its results obtained is the difference between the public and private sector, with the public sector having an autocratic leadership with poor management autonomy (Dominguez, 2008)?

In Veracruz, a diagnostic study was conducted to determine the leadership profile of the manager and his collaborators in the operational management of a chemical company, using the model range Bass and Avolio. It uses the instruments proposed by these authors as multifactor leadership questionnaire. Among the findings highlighted there is an organizational and preponderantly transformational leadership, and to analyze the dominant leadership behaviors dominated by a contingent reward (Garcia, 2013).

METHODS

To obtain data of first level of development an interview with the director was taken and the project scope was established. Below is a guide of questions that was conducted to later meet with key personnel at all levels: administrative and operational. The sample was determined on the criterion of discrimination the years of seniority: less than 1 year, 3 to 5 years and 10 years and older. Subsequently a non - participant observation was conducted to gain insight into the working environment and identify their most obvious characteristics.

An adaptation proposed by Hersey and Blanchard to determine the situational leadership style instrument was applied. The sample size was 10 people, which represented the entire management and control and were classified as follows: managers and area managers. A work plan was agreed with the CEO, same as revised and suggested amendments to facilitate understanding of the questionnaires and protect the confidentiality of respondents, preserving the original format to meet the desired objective. To all personnel it was applied the instrument 213 NTP: Job satisfaction: evaluation survey from the Minister of Labor and Social Affairs of Spain, to determine the perception of workers about their work and providing clues about the motivational profile of subjects.

With the data obtained, the diagnostic tool called Pareto chart was applied to identify the main causes generating this problem.

ANALYSIS OF RESULTS

As a result of the interviews, observation and surveys of all staff, then the characteristics of the current situation of the company are listed:

- Workers are limited to doing what is asked of them and not go beyond, that is, they are more reactive than proactive.
- Some employees both in key positions and technical positions are in a state of comfort, because they know that regardless of the results they receive their salary and bonuses.
- Employees do not put on the company shirt, do not seek the common good, and only care to see what are given to them.
- For employees, it is not clear or easy to understand the rules and guidelines of the company for perceived injustices among staff.
- In the offices, it is common for employees misunderstand and abuse the freedom and confidence of the leadership style of the company.
- In the daily work, there are employees who do not respect the work schedule and company standards because the sanctions are not executed.
- The leadership has not been achieved in some cases because they take away authority (the manager decides and another senior officer arrives and says otherwise)
- It is perceived a permissive authoritarian leader and one, each with its language and perceived actions style.

Then the perceptions of the company employees are listed:

- GISITCA is a company with competitive advantage; specialized work is done with the latest technology, so you have a status and constant challenges.
- There is willingness to learn and research to meet the projects.
- Projects are different and this makes the job interesting.
- The owners are very human and good people (patronizing).
- The technical ability of employees is from good to very good.
- There is willingness to learn and research to meet the projects.
- Employees perceive the company competitive advantage for specialized projects made with advanced technology.
- The projects are different and this makes the job interesting.

After collecting the data, they are grouped by category and the presence of each type of problem is recorded in an orderly table as a basis for Pareto chart (table 2).

Later, totals are plotted and cumulative percentages resulting in Figure 5.

The information provided by the Pareto chart indicates that the 3 main problems to be solved are no clear guidelines, lack of teamwork and not effective leadership.

As a result of applying the questionnaire on leadership style Hersey and Blanchard to senior management, the following information was obtained (table 3):

As shown in Figure 6, the directing is self - considered mainly authoritarian but open to support all processes, i.e. a paternalistic leadership where there is a low delegation of authority and therefore low support to facilitate decision - making by collaborator.

Intervention Plan

Taking as a basis, the 3 main problems identified in the Pareto chart, it can be appreciated that 2 of them are of the same nature: leadership ineffective and lack of teamwork. So applying effective leadership empowerment, are encouraged to

Table 2. Cumulative for the Pareto Chart

Problematic	Events	%	% Accumulate
Unclear guidelines	40	25.6	25.64
Lack of teamwork	39	25.0	50.64
Ineffective leadership	37	23.7	74.36
Reactive personal	28	17.5	92.31
No sanctions	12	7.69	100.00
TOTAL	*156*		

Figure 5. Totals and cumulative percentage
Source: Own Elaboration

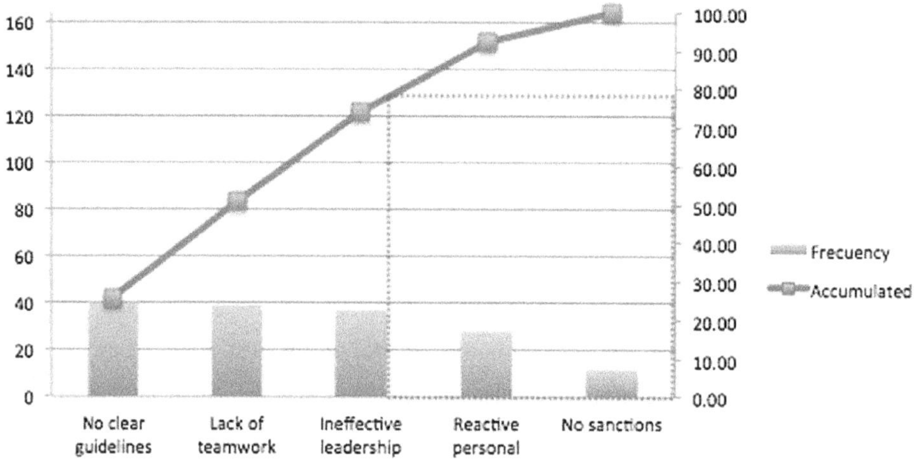

Table 3. Auto appreciation of leadership style

DIRECTING		COACHING		SUPPORTING		DELEGATING	
2	4	1	3	4	3	3	2
7	3	6	2	5	4	8	3
11	3	10	1	12	4	9	1
13	4	15	4	14	3	16	2
17	2	19	1	20	2	18	2
23	3	22	3	24	3	21	1
28	4	25	1	27	2	26	2
29	4	30	2	31	4	32	1
35	3	34	1	33	2	36	1
40	3	39	3	37	3	38	2
41	4	43	1	42	3	44	1
48	3	46	1	45	3	47	1
TOTALS	40		23		36		19
Characteristics							
Directs Controls Own plans Instructs Supervises Monitors		Coaches Guides Clarifies tasks Offers advice Gets team ownership		Supports Consults Outlines task Discusses Invites ideas Gets team agreement		Delegates Facilitates Gives overall direction only Gives team responsibility Expects team to report on progress	

Figure 6. Preferred leadership styles today
Source: Own Elaboration

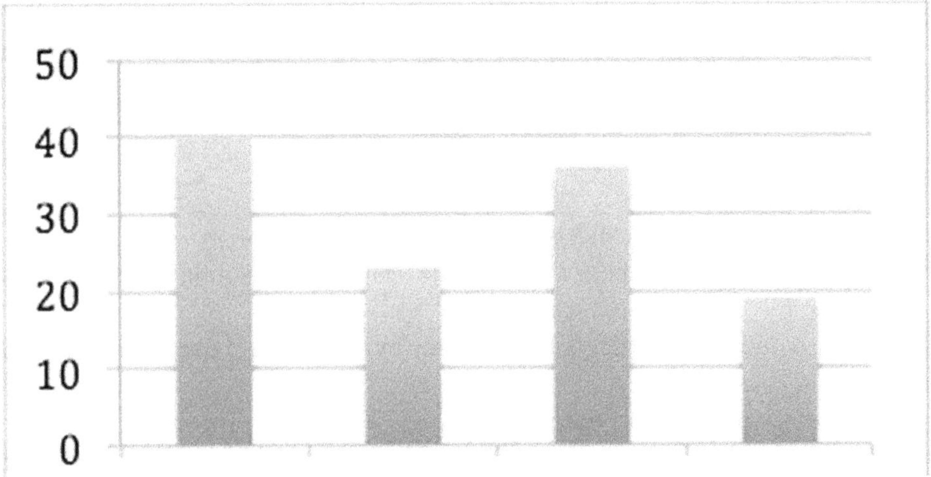

employees and teams to execute orders and improve processes, products and services with positive results in growth and organizational development. The remaining problem about no clear guidelines is addressed by performing revision and updating of administrative manuals: organizational manual, manual of rules and procedures manual positions and functions, the training manual and also apply techniques for dissemination and training to give know the updates.

1. **Evaluation System:** The other important aspect for the organization to achieve the expected success is to improve business productivity, keeping its staff motivated, for which a system of results evaluation should be established through key process indicators (KPIS). It is suggested the creation of a Balance Scorecard by department, taking into account the following aspects:
    a. Update the client - provider diagram of the organization.
    b. Each department should establish its productivity indicators with its respective internal customer.
    c. Filling and application of the Balance Scorecard which will show an overview of the general state of the organization, from the financial perspectives, customer perspective, perspective of internal processes and the perspective of learning.
    d. With established indicators, individual and group performance will be measured.

2. **Incentive System:** In order to establish the incentive system, the following criteria must be considered:
   a. Management should consider a budget program for this purpose.
   b. The management should establish what formal incentives to give, when to deliver them and determine the number of workers benefited.

   In the case of GISITCA it is suggested to consider the following:

a. The incentive will be by levels according to the increase of the monthly productivity.

b. Incentive for punctuality and monthly assistance.

c. Recognition of the best worker of the year by maintaining the best results and granting recognition and a bonus.

d. Recognition for loyalty of the company by remaining active over 10, 15 and 20 years of service.

e. Management should encourage supervisors and area managers to make informal recognitions to workers who demonstrate self - improvement, effort, and dedication to their activities.

Here in Table 4, the areas covered by the intervention plan detailed.

Table 4. Intervention plan

Problem	Proposed Solution
Unclear Guidelines	• Review and update of administrative Manuals: • Organizational manual • Manual of rules and procedures: • Incentives • Permits • Reviewing manuals • Manual of posts and functions • Training manual • Dissemination and communication upgrades
Ineffective Leadership	Courses of leadership, direction and empowerment to senior management.
Lack of Teamwork	Empowerment ▪ All staff: Teamwork course Conflict management Technical courses according to the area Middle Management: Comprehensive development courses ▪ Middle Management ▪ Course management skills ▪ E4 Partners: empowerment courses ▪ Performance Evaluation System ▪ Balance Scorecard (performance indicators) ▪ Improvement of Results ▪ Incentive system
Performance Evaluation System	Balance Scorecard (Performance indicators)
Improvement of Results	Incentive systems

CONCLUSION

In this paper, it has been analyzed a very common type of leadership in Family SMEs in growth, with a paternalistic style, and the problems this creates in the workplace. The characteristics of the theoretical model most used in industry leadership for analysis were presented: Situational Leadership II Blanchard. With regard to BPA companies in the electricity sector, these are facing a part that is opening up the sector, promoted by the energy reform in Mexico. With this the doors are open to reach power companies that require services of these businesses, and is also expected strong domestic and foreign competition from providers of the same service, especially Chinese companies.

In this essay, a guide is provided for SMEs that are growing to migrate from the paternalistic leadership with which they were created to democratic leadership that enables them to successfully meet the challenges ahead, with that as stated by Blanchard (2007) if to provide better customer service, the employee needs to be motivated and for this, it is essential to treat it well, empowerment is the solution. To achieve this, it is needed a comprehensive and flexible manager who are able to adapt their leadership to the situation and to extract the best from each employee.

Because of the potential showing to have a good leader as a director of the company who can motivate and support his employees, it is recommended to generate future implementation plan, carry out and evaluate it to determine the achievements.

REFERENCES

Arias, F., Basualdo, L., & Heredia, V. (1989). Performance measurement. In F. Arias (Ed.), *Human Resources Administration (4th ed.)*. México: Trillas.

Belausteguigoitia Rius, I. (2003). *Family businesses. Its dynamics, balance and consolidation*. México: McGraw - Hill.

Benavides, E., & Delgado, Y. L. (2004, May). Analysis of situational leadership and disposition towards tasks based on skill and ability. In *Memories of the VI International Congress of Ergonomics,* Universidad de Guanajuato.

Blanchard, K. (2007). *Leadership at the highest level. How to Create and Lead High Performance Organizations*. México: Editorial Norma.

Blanchard, K., Zigarmi, P., Zigarmi, D., & Halsey, V. (2013). *Situational Leadership II*. The Ken Blanchar Companies. Retrieved February 12, 2016, from http://images.leadership.kenblanchard.com/Web/TheKenBlanchardCompanies/%7B26e2fdd1-daa7-4b13-be24-752d41d34a1b%7D_The-SLII-Experience-Sample-PW.pdf

CFE. (2012). *Electricity sector works and investments program 2012 - 2016.* México: CFE.

Chiavenato, I. (2000). *Human resources management.* Bogotá, Colombia: Editorial Mc Graw –Hill.

Chiavenato, I. (2002). *Human Talent Management.* Bogota, Colombia: Editorial McGraw-Hill.

CONDUSEF. (2013). *Common mistakes of the SME entrepreneur.* Retrieved from http://www.condusef.gob.mx/index.php/empresario-pyme-como-usuarios-de-servicios-financieros/548-errores-mas-comunes-del-empresario-pyme

Consultores, A. (2016). *Pareto Diagram - Quality Tools.* Retrieved from http://www.aiteco.com/diagrama-de-pareto

Cortes, J. (1999). Factors of leadership and competitiveness in a financial services company. *Revista ABANTE, 2*(1), 71–96.

Delgado, N., & Delgado, D. (2003). The leader and leadership: Reflections. *Interamerican Journal of Library Science, 2*(26), 75–88.

Diario Oficial de la Federación de México. (2013). *Decree amending and adding various provisions of the Political Constitution of the United Mexican States, in Energy Matters.* Retrieved from http://www.dof.gob.mx/nota_detalle.php?codigo=5327463&fecha=20/12/2013

Entrepreneur, S. (2011). *How to do empowerment in your company.* Retrieved from http://www.soyentrepreneur.com/como-hacer-empowerment-en-tu-empresa.htm

García, J. (2013). *Diagnostic study to determine the leadership profile of the manager and his collaborators in the operational management, according to the full range model of Bass and Avolio: Case of a chemical industrial company in Coatzacoalcos.* Veracruz, México: Revista Caribeña de Ciencias Sociales.

García, M., Posada, V., & Hernández, C. (2012). *Motivation and reward systems and their impact on production, Contributions to the Economy.* Retrieved from http://www.eumed.net/ce/2012/gvr.html

INEGI. (2015). Business life expectancy. *Boletín de prensa, 87*(15), 1 - 2.

Instituto nacional de seguridad e higiene en el trabajo. (1998). NTP 213: Job satisfaction: evaluation survey. Retrieved from http://www.insht.es/InshtWeb/Contenidos/Documentacion/FichasTecnicas/NTP/Ficheros/201a300/ntp_213.pdf

Kaplan, R., & Norton, D. (2005). The Balance Scorecard: Measures that Drive Performance. *Harvard Business Review America Latina, 83*(7), 102–110.

Lee, J. (2004). The Effects of Family Ownership and Management on Firm Performance. *SAM Advanced Management Journal, 69*(4), 46–53.

López, J. (2005). Labor Motivation and Human Resource Management and the Theory of Frederick Herzberg. *Gestión en el Tercer Milenio, 8*(15), 25–36.

Maslow, A. (1943). A theory of motivation. *Psychological Review, 50*(4), 370–396. doi:10.1037/h0054346

Nelson, B. (1996). 1001 ways to reward employees. Colombia: Norma.

Peiró. (2000). Psychology of the Organization. *Universidad Nacional de Educación a Distancia, Madrid, 159*, 165–170.

PROMEXICO. (n. d.). *PYMES is a fundamental link for the progress of Mexico.* Retrieved from http://www.promexico.gob.mx/negocios-internacionales/pymes-eslabon-fundamental-para-el-crecimiento-en-mexico.html

Real Academia Española. (2017). *Diccionario de la lengua española.* Retrieved from http://dle.rae.es/?id=aBuhX28

Rodríguez, D. (2006). *Organizational Diagnosis.* México: Alfa Omega.

Sánchez, E., & Rodríguez, A. (2010). 40 years of situational leadership theory: A review. *Revista Latinoamericana de Psicología, 42*(1), 29–39.

Tannenbaum, R., & Schmidt, W. H. (1973). How to choose a leadership pattern. *Harvard Business Review, 51*(3), 162–173.

Yukl, G. H. (1994). Leadership in Organizations (3th ed.). Englewood Cliffs: Prentice - Hall.

KEY TERMS AND DEFINITIONS

Business Process Automation (BPA): Business Process Automation (BPA) is the strategy implemented by a business to automate operating and controlling systems, methods, processes and techniques by integrating software applications and electronic devices in order to reduce labor resources, human intervention and costs.

Incentive System: An incentive system is a formal business management program and tool that introduces a structured motivation scheme to motivate, promote and encourage specific desired behaviors and actions of group people in organizations to achieve certain goals during a period of time.

Organizational Climate: It refers to the conditions and properties of the business environment and within an organization in a workplace, such as the staff practices in motivation, rewards, leadership, communication, conflict, etc., that influence organizational performance.

Organizational Intervention: It is a method, technique and process of structured activities used in organizational development to improve its social and task performance.

Planning: Planning is a management process in organizations, concerned with determining the mission, formulating strategies and policies, defining goals and establishing resources to achieve targets for future direction.

Situational Leadership: Situational leadership refers to the leadership style adjustment of the leader that may change continually based on the specific situation to fit the development level, meet the needs and influence the followers in an organization.

SMEs: Small and medium enterprises.

Chapter 6
Digitalization's Impact on Work Culture

Manisha Mathur
National Law University, India

ABSTRACT

Digital work culture followed by any organization on the digital workplace in digital environment, work is going through the technologies in the virtual office, business site merely presents in cyberspace by using their laptop, cell phones via accessing the internet, anywhere, anytime. It directly affects economic, social, culture and business over the globe in a nano-second equally. It plays an important role in the development and growth of any organizations as well as the economic growth of the nation work culture is actually beliefs, ideas, manner and attitude of employees along with organization's ideologies and principles. author presents information about Digital media has impacting on Work Culture now traditional work culture is become digital work culture where efficiency with qualitative and quantitative productivity of employees and organization and business in a very swiftly, time saving, economically by the sharing information globally.

INTRODUCTION

This chapter studies impact of digital media on work culture and the contribution of advanced digital technologies and the devices in efficiency and productivity of employees and organizations. The positive and negative aspects of digital media, effective handling of digitalization and its grievances are analysed. On the working place, digital media offer an interesting working environment and provides an interactive platform for smooth communication between employees and management.

DOI: 10.4018/978-1-5225-3191-3.ch006

Digital media is ready to lend a hand to develop better understanding between employees and higher management in the organization. Digitalization helps in developing a positive work culture which in turn enhances the productivity of employees and finally results in the development of organization's business globally. The attributes of digital media enable global information sharing, saving of time and swift business transformation.

BACKGROUND

In this digital age, knowledge and information spread amazingly fast and at a mind-boggling speed. It directly affects economic, social, culture and business over the globe in a second equally, impacting all. No doubt, digital influence of company extending in global market helping the organization to perform smoothly. It is also improving efficiency and quality of their products with day-to-day operations. The prompt innovation of technologies is with the motives that allowing digital contents audio, video and graphics contents transmitted over the internet by the digital media like, window files help employees interact globally and make them professionally quicker and most efficient, which was not possible in earlier times.

Artificial intelligence is a powerful feature of digitization that can be automated the work like human being. It has a great supporting feature on various aspects to track and handle the functions of employees and various aspects of their job. It helps in developing healthy work culture and makes the staff more competent as demanding by employer's expectations and also supportive of decision making to take the solution of a specific area of the problem by the experts of relative fields.

Online Banking provides facility to check account, easy transfer money, etc. Online selling and purchasing, e-payment system facility like e-cash, e-billing, e-wallet, smart card, etc. are facilities provided in the digital environment.

It is important to understand that digital media could play a crucial role in shaping the development and growth landscape of any organization. In this Digi-tech age, digital media is affecting work culture in organization. It is transforming work succession rate into a digital setting of success. Digital Media has played a significant role in any organization to maintain a smart work culture.

1. **Work Culture:** Work culture is defined as a unique social, psychological environment including values and behavior of an organization and its employees. It plays an important role in the development and growth of any organization similarly economic growth of the nation as well. If work culture is friendly and healthy, the employees will be concentrating on their work with job satisfaction, and an aura of positivity will prevail all the time. Employees will remain in

the organization with the qualitative and quantitative product. Management can perform its functions swiftly, smoothly and with great achievements. That said, work culture is nothing except the mentality of the employees which further rectify the environment of the organization. Work culture is beliefs, ideas, manner and attitude of employees along with organization's ideologies and principles.

2. **Digital Work Culture:** Digital work culture is a culture followed by any organization on the digital workplace. Digital culture provides a good and healthy work culture which is important for the growth of any business as depends on efficiency and satisfaction of employees. A healthy mind, healthy body and healthy spirit have a healthy worker and healthy thinker and a good or healthy work culture is helpful in maintain satisfactory interpersonal relationships. Network maintains user friendly atmosphere in which leaders can trust and a sense of self-empowerment. Networks guide management towards visionary task that generates innovations and creativity.

For academicians, digital library plays a significant role. It works as a pooled. Printing and storing the material as well as for library building could do with less space. It creates digital databases for storing, retrieving, updating the information and knowledge. The way, it saves the space and provides further space for group study. If you have internet connection on your digital device can access it everywhere and anytime. Students, faculties, and staff are enjoying their work on digital environment and encourage for increasing and sharing their knowledge along with work by the administration through maintain and foster the secure digital library. Now, digital world offers efficient, creative and economical, easy working environment, and projects for employees as well as the organization and also ready to lend a hand selling and purchasing company products.

Employees are enjoying their work and also enforced to spend more energy to accomplish the task which they can do with much lesser efforts in digital work culture. Employees and management both will be disciplined and co-operative, provide an environment for encouraging a healthy discussion between them. As a consequence, unity establishes between all members and employees and they work as a team member. Hence, employee satisfaction, motivation will be high, and they are ready to give the best performance without any problem.

Work performance and behavior are constantly affected by physical and digital conditions of work. If working conditions are not proper, increasingly realized so many work problems and psychology related such as low morale, absenteeism, accidents, and fatigue, etc. Working hours are too long creates monotony, feeling of tiredness, as a result it leads to accidents, low and faulty production. Consequences decline in quantity and quality.

3. **Interpersonal Relationship:** Between employees and management plays a significant role in the development and progression of any organization. Digitization provides facility to office automation electronically, where an employee uses management software in the e-form (electronic form) to create, integrate and transform the forms and form data with processing systems along with portable documents format like.pdf files and for dealing out documents and to capture data for pre-filing provider database also.

Employee's mindset concerned with indifference, emotional instability, hostility, frustration and a swarm of other such erratic that may lead to accidents. Employees need safe, swiftly and hassle-free working conditions.

It helps to get higher productivity with great improvement, better relations with customers and clients and increasing flow of communication. The organizations need digitization for both healthy behaviour and inspire superior performance and provide satisfaction. Various psychological and socio-economic factors have effects on developing individual's attitude positively as well as negatively towards members of the organization.

4. **Digital Work Place:** Digital workplace is the environment where work is going through the technologies in the virtual office, business site merely presents in cyberspace by using their laptop, cell phones via accessing the internet, anywhere, anytime. The invention of internet and web 2.0 technologies create an educational, social entertainment for every nook and corner of the earth.

The use of technologies in the working place becomes the smart workplace for the both employees and organization. It provides information on demand or target. Digital work culture encourages a corporate culture of autonomy, accountability, and empowerment of employees.

Digital Revolution

Third Industrial Revolution (Humbert, 2007) has an information revolution rising of network society which commences of up to date social, economic, technological trends known as digital revolution based on digits (Rifkin, 2011). Commencement of omnipresent full efficiency, cost-effective use of analog format like cassette tape is converted following bit-by-bit swapping into a popular favorite approach like optical discs *digital format*. The digital era has been started during 1980's as personal computers and the Internet along with devices exist into the market while the Industrial digital revolution begins from 1771-80 based on the water-powered mechanism. By the innovation and commencing new digital communication technology people are interacting and connecting with each other every nook and corner globally, their knowledge increases and transformed easily and promptly.

In these steps, digital revolution, dominant technology (Sterling, 2009), cloud computing technology appeared in 1996, refers to platforms for distributed computing with the object to facilitate the user's spotlight on their heart of the business. Cloud computing technology with time sharing and open source cloud software invented digital devices like cell phones, smartphones and tablet computers.

iPod portable media player designed and marketed by Apple was launched on October 23, 2001. Market shares of PC Vendors, first commercially successful personal computer to feature a mouse and a graphical user interface MAC or Macintosh was introduced by Steve Jobs CEO of Apple Inc. on January 24, 1984. iPhone line of internet and multimedia-enable smartphones designed a marked Apple Inc. Originally released in 2007. An iPhone function as a cell phone (also including text messaging and visual voicemail), a portable media player and an Internet client (with e-mail, web browsing, and Wi-Fi connectivity). The user interface is built around the device's multi-touch screen, including a virtual keyboard. Third-party applications are available from the App Store, which launched in mid-2008.

Digital Revolution in India

Digital era was commencing from 20th Century and biometrics established in 2002, for real life attendance developed by Mantra Softech Pvt. Ltd. Fingerprint, smart card, face-reading, etc. digital. It shows that it was the first step taken by the Indian government to commencing of digital India and now, our Prime Minister Mr. Narendra Modi has been taken a very good step digital payment for everything by every person implementing towards digital India.

A digital media receiver can connect to the home network using either a wireless (IEEE 802.11 a, b, g, and n) or wired (Ethernet) connection. A Digital media receiver includes a user interface that allows users to navigate through their digital media library, search and play back media files.

MAIN FOCUS OF THE CHAPTER

Digitalization and Changing Work Culture Technological Changes

The innovation of digital technologies and devices has a great impact on work culture. A great revolution comes into the work culture, its turn out to be revolutionizing on the ground of working knowledge, efficiency, speed, quality, and quantity is increased, and many more people are enjoying their work with their efficiency, in this digital era, employees understand the nature of their work, market has been increased. Globalization of online marketing and its high rates shows the economy of market growth and global trends further information, acquaintance, realization, and wisdom produce last five years as the times gone by and assumed that further knowledge would be double in every five years.

Digital revolution has been increasing revenue, reduced costs with the progressive employee and high productivity, enhancing the environment, safety, security and citizen's reality, healthier and well-being in the public as well as private sector.

In style, the social media for a reason give the go-ahead to collaboration and creativity. The contribution of the audience of the social media by including comments or modifying the stories and the empowerment and freedom of audience to add or create new content makes social media a process and not a static medium.

Issues, Controversies, Problems

Privacy and Piracy

Privacy is a major concern area of managing the security of information. It is right to privacy of an individual and support to sustain and look after the confidentiality of information of individual and organization. Confidentiality is one part of privacy covered by the right to privacy. Right is not properly identified by the most of the country's constitution all over the world. But in General term, but in common sense, privacy defined as a liberty from illegitimate interruptions.

Piracy is unauthorized downloading of material, modify the material for disturbing people send online or digitally in contemptibly by using P2P (peer-to-peer) technology. "Bit Torrent" is the best example. The Internet is the rich source of information provided by the websites, online end users; openly provide personal information also using the software on the registration page, order form, survey, and so many ways. Cookies are helpless to directly find out the visitors name and addresses and e-mail embedded graphical files called Web bugs along with web pages originated monitor for catching email messages readers and the web page with the information is transmitted to the another computer and some other secretly installed illegal software on internet user's computer through the piggybacking on the large applications are creating violations on the user's computer and harmful for maintaining privacy on system. Sniffers used by the activist for the purpose of damages the system with attacks on the privacy for stealing proprietary information and complicated to detect, such as e-mail messages, company files, and confidential reports. All are commercially eye-catching for business to demeanor observation on a gigantic degree seeing that realize growingly new ways know how to use personal information to acquire benefits in the marketplace. Extremely vulnerable data are collected to the business for the large market product and services as well as permit boost the revenue by selling and purchase on the cyberspace. More and more browsing or surfing is not only for purchasing or selling the goods but also for gaining the personal information. Hence, the loss of privacy and confidentiality of a person or leakage of vital & confidential information any organization is a major issue come out in the digitization environment. Information or information contains access risk of exposing personal data, behavior related information and knowledge also an alarming issue.

Different Perception of Countries for the Data Privacy: Data protection under UK Law defined Section 15 (2) and section 7 (9) of the Data Protection Act, 1999. US provide data information privacy by Federal Trade Commission to the adult and to the children by Children's Online Privacy Protection Act (COPPA) in October 1998. Indian Constitution provides privacy law under Article 21 whereas computer related privacy issues are dealing under the section 43, 65, 66, 70 & 72 of IT Act 2000 and the Indian Contract Act 1872 afterward the Protection Data Bill was introduced in 2006 under Indian Penal Code with the aim to give the protection of personal information of the people. Whereas the section 69 of IT Act 2000 of India describes the ground under which the government interferes with data; this section includes both interception and monitoring along with decryption for the purpose of investigating cyber crimes. (Kamath, 2012)

Consumer Behaviour

Consumer behavior becomes more dynamic and unpredictable that effects on sell and create pressure for continuous innovation. Consumers gain the knowledge about the product in a very easy and quickly by the digital media, hence they are more knowledgeable and now their requirements, expectations are frequently changing, and new demands are promptly raised. It is very beneficial for the growth of marketing by the competitiveness, demanding innovation and creativeness but market pressure is also increased.

Business and Organization

In the business and organization are in nature becomes more strategic and competitive because of innovation and creativity in every aspect of demand. Consequently, eventually, risk exposure has increased.

Stress for Work

Digitization creates a competitive environment, which demands continuous improvement in skills and people works 24*7 hours means unlimited working hours which effects on individual health, diminished social relations, etc.

Deprive of Humanisation

The living standard is going very high, technology enable skills and competencies are in demand, job opportunities are reduced in some way, time spending on learning new technology. Everywhere process of automation seen as a result can be seen affected and target the process of humanization.

Digital Felony

Digital workplace has a great risk. Powerful easy to use criminal tools are interrupting by the secretive forums. Some bright minded Web host and individual specialists, engineers, are very creative, and tactics using engineering threaten called web engineering becomes criminals they hack corporate database using the fully fledged infrastructure of malicious code writers cleverly they hire teams of networks of thousands of computers to transmit automated attacks. That said, they hack computer by the malicious software first, and then they do whatever they want, even they hack organization plans from foreign IP address. Using the cookies (small text files stored by the server) activist without knowing the user track user's behavior and data. These cookies can store viewed information such as user's name, password, age, sex and other personal data for the purpose of creating profile for misusing the original profile of user, digital video recording, Today, widely spreading enormous issues on the work culture relating to digital felonies such as Identity thefts, malicious programmes are flowing secretly, personal sensitive along confidential and financial data is stolen, sexual harassments like:, fake signatures of authorities, cybersquatting, ATM or net banking frauds, Money laundering, fake accounts creating on social media. Sometimes person is not hacking your system, but they can make duplicate account of your profile to the acquire money or wanted to defame you as well as your Institution or company, etc.

Geographical boundary-less environment and nation's different culture is a hurdle to handle cyber felony all over the globe. To make common law for all the nations is a critical aspect of fighting against well-known white collar crimes "cyber-crimes" or "digital crimes." By the hacking, your or government official sites activist use official confidential data or document for their purpose by this way offenders are creating digital combat environment on the international level terrorism. Even the different countries have different laws against those crimes or offenders but the nature of digital technology as evidence helpless to prove the crimes and criminals.

FUTURE RESEARCH DIRECTIONS

Future and Emerging Trends

Big data and Open sources are very popular these days. Business and the digital market become easier, faster, comfortable, with greater productivity and greater profit over the entire globe through the SAAS, PAAS & IAAS are the cloud computing elements. The main emerging trends are followings -

1. **Embedded Data Analytical Technologies:** Invisible and real-time automated, easy use tools perfectly availability with more pervasive intention for the moment of decision making, and during the own daily workflow and task doing the business users and customers. All kind of applications and users can access these tools for the data analysis and business intelligence. Tools are SAP, SAAS, LogiAnalytics, IBM, etc.
2. **Production Studio Technologies:** Studio technology was designed and built in U.S. for the use World Wide Web broadcasting and serve as an audio control hub for announcers and commentators. Production Studio Technology facilities to package and deliver the content in a safekeeping or a novel manner of sales or internal presentation through the video, infographics, podcasts and next generation presentation software to the employee.
3. **Personal Cloud:** Personal clouds are the devices which employee or customer use according to their preference like Apps, devices, and things. Quickness, awareness, and engagement other than security, complaints, and compatibility is the concerning area of the personal cloud. This can be controlled by the managed diversity approach that assurance to the organization on the right balances between employee autonomy and control.
4. **Virtual Personal Assistant:** These tools make our daily work and divergent viewpoint of the business easy and Fast with relatively low cost. As well as the technology improves, employee grows and feels comfortable with the outside of work. Customarily VPA is being used for data retrieval and basic task. Assistant tool and techniques are Google now, Echo, Siri, freelancer.com, virtual staff finder, etc.

5. **Security Chatbot Net:** Chatbot is an open source and attached with detecting and alerting system of drop box known as security bot to tackle unauthorized access and other security incidents in the workplace. It has a feature of automatically grabbing alerts from security monitoring tools and verifies incidents with other employees. The bot is mainly built for slack and designed opt for transferable to another podium also. This collaboration channel is consequently boosting the efficiency of communicating day by day employees' queries and security alerts.

In the future, the everyday job will be completed by the digitally, and the risk of intrusions emerge through the digitally in the organization, its effect will be shown in the business by digital trends as a massive moderate level. So, the function of digital media becomes more important and critical. To give the direction of employees helpful to give stability of employees among work and their personal life-related activities and for fine tuning in their modifying role near future research is required.

Development in the Indian Scenario

Digital India a vision of Prime Minister Modi is a deliberating give a ring to welcome steps of online payment towards the great opportunity for declaration the India in the third Industrial Revolution before that Information and Communication Technologies (ICTs) has never been a greater exploit.

Digital media has dramatically increased productivity and creativity by the competent. The real community will have been renovating into the cyber community by the web-based power in near future and destiny will be so divergent, the world has been disparate and will change ever faster information, collaboration and knowledge sharing are essential organization secure opportunities for their future, flexibility is the success factor in the transforming world. Radical transformation new form of organization fundamental changes come to pass in every week, day, even hours, situation impacts many and more people want information, social network friends are more trusted than traditional media and government, continuous improvement and customer dialog.

Virtually the digital way of life is an entirely new trend of exploit business and for work culture of any organization.

Digital Technology and Impact on the Office Work Culture

Digital technologies have a positive impact on work culture. Digital communication is the moral fiber of any organization by the use of many applications and systems headed for the contribution of information conveying the messages, accumulate information to download by the success the customer. Company's information can be shared with selected addresses in the live mode via the digital presentation through the boards, flip charts, over projectors and transparencies, power point presentations including pictures, images and matter with automatic and simple attractive manner. This visual gripping and impressive impact on the visual aids becomes more widespread with the laptop and more supportive for the company to motivate their employees to achieve the aspiration and to attract the customers. Digital communication provides an interesting over and above effective communication system for document sharing facility to store, update, delete and search emails users according to their needs. By the video or electronic devices conferences and meetings handling becomes very easy and more comfortable, now voice conferencing is also emerging. Real-time virtual communication is going through speaker phones or network PC's that have telephone network or internet telephone connectivity; video conference offers the capabilities of video and audio for different locality partakers also. A great feature of digital media is web publishing using websites and portals for storing documents, catalogs, drawing a picture and so on for sharing information stored on the documents, the user can search, navigate, select and downloads documents for self-use. Web publishing is more popular among researchers, educationist, large business, commercial bodies, government bodies also popular to share information with communities. Digital communication system offers their users to send messages, documents, and files in any format over the internet.

Enterprise information gateway the web-based interface tool on an integrated internet/intranet/extranet platform is allowing customers to use application and other services for posting the information and communication. (Jawadekar, 2006)

The collaboration of virtual team with different members are virtually working with electronic technologies along with network or the internet for communication, they access different databases and the servers to successfully accomplishing the assignments and achieving the common objectives of the organization. These technologies along with the internet and network control devices the communal wisdom, acquaintance, skills, and familiarity of the limbs.

Work flow through the digital resources: Groupware software or communication handling best tool for the step by step work perform by using the internet is increased its effectiveness by the allowing members to sharing information and common database, invoke an application and work together to create and share the document and so on, such as Novell GroupWare, Microsoft Exchange, Lotus and Netscape Communicator are GroupWare tools.

More profit created by the high spirit employees because they are more efficient and more productive. In high spirit, workers have a positive feeling which insists wits a great job radically persistently. Consequently, employees are effectively working towards the organizational goal. Own contentment of employees has augmented them makes confident and good leader possible. Eventually, effects can be seen as improving productivity and increases profits. Digital applications along with digital devices like cell phone and internet act as a meaningful role in the communication and information tools amongst the online workers. Digital Information Communication widely spread the enormous population outer surface of the company, provide flexibility in their working hours and they do their works for long hours. Workers are having an interest in digital communication. Therefore, the excess use of email like digital media as a key to their jobs, using such type of media and other social media continues to be main digital channels are emergent potential threads like; phishing, hacking and spam, etc. and these are disastrous cautious about productivity lost and misses of email. And office-based occupations are critical for online workers. Sometimes digital media may be a distraction in the workplace.

To give fabulous productivity and keep up to date with competitors' social media plays an important role by the tracking company's productivity and true way handling. Desktop creates transparent, accountable environment during working of the employee, hence managers and employees have their mindset to identify global trends by the way productivity improved.

In Digital Era, worldwide workers are promptly increasing linking through the mobile computing connected by the data as well as voice and the virtual connection between the administration and their working parties via Skype, VoIP, email, and cloud-stored files are helpful for workforces autonomous to go anywhere and do their work freely, comfortably with full efficiently instead of their original working place.

For the employee and organization skills development is making upright progress easier by the easier access using digital resources. Innovators, technical experts, and experienced professionals from over the globe collaborate jointly digitally integrated "on-demand" working parties can be created by the company's that offers benefits hooked on wide-range of networks. Digital system is also supportive of making a balance between all revolutionize and implantations of the needs of workers as well as companies.

Digitalization and Office Work Efficiency and Productivity

Digital culture is more supportive to understand the nature of their jobs and to change the perception towards jobs and the organization. Employee work in digital culture will be satisfied, motivated, ready to give best performance without any problem because digital work culture improve, increase human being productivity, connectivity, working speed, communication skills and changing work ethics, real-time connectivity, create and maintained friendly relationship between employees and management, employees and management both would be disciplined and cooperative, they work as a team member, encourage discussion, unity between all members and employees. In fact, the brain works significantly well by the positive feeling. Digital culture intensifies positive feeling and own happiness in employees for effectively working towards the organizational goal and creates more profit. They have confidence in their leadership and may be a good leader. The ultimate result productivity and profit will be a boost. Email, cell phones, the internet, etc. all digital media plays a significant role in the communication and information tools amongst the online workers.

Digital Media is the combination of various digital technologies that increase interest and confidence of employees towards their jobs. It is to see in the mind's eye, the possibilities on the work culture business manager clear that fetches probable improvements and benefits like peas on the pod to the employees realize. Employees' confidence is encouraging, so they are optimist towards their job and digital technology. Feasibility of working place is persisting to the software engineers to build advanced technologies and obvious business.

Possibilities of workplace re-invent technologies time to time business leaders and employees understand same possible benefit digital technologies can fetch. Using and accessing digital collaboration tools on the workplace employees feel more satisfaction and most of the time they are energized to do work.

Technology has a crucial role to play in enabling organizations to create an environment that sustains the goals of their employees, which is to work in an environment with a good workplace. It creates new and more rapidly tactic for businesses to access data, talent, and customers. It provides customers with access to new products, information, and services that improve their lives. Social media can be useful in keeping up to date with competitors and gives fabulous/excellent productivity. When employees are working and monitoring on the desktop it creates transparent, accountable environment. Employees and managers have their mindset to identify global trends and way to improve the productivity. Current time, the digital world provides low-cost easy, creative and efficient working environment and projects for the workers as well as companies. These projects are helpful to selling and purchasing company products.

SOLUTIONS AND RECOMMENDATIONS

Digitalization and Grievances Handling

Every invention has cons and pros. Digital media is also not untouched; some pros are assisted positive work culture into negative.

To track organization and company's productivity in a true way handling websites and some useful applications are available in the market.

Employees do not understand how criminals are working and how they can be targeted. Authorized anti-virus software should be installed by the company or organization for giving data security time to time on every system. The employee must be permitted to access only those data systems which they required for the work only and company or organization should not be allowed to install unauthorized software on the company's system by anyone even employees, authorities also. It should be taken as a disciplinary.

Open to Innovation and Adopting New Digital Methods

Company should be built up the new ways or schemes of digital etiquette time to time according to company's requirement and employees expectations. Some office suite related authorized software available in the market, as the requirement of organization or company would be purchased or employee allows to free access them, such as free Accessible Google drive doc, one drive cloud services, Hyper office, free, open-source replacement for Microsoft Office openoffice.org Office Suite. (Duffy, 2014)

Online Meeting or Discussion

When the online meeting is going through the productivity tool like WebEx, the setting of this tool should be one day previously before the actual meeting will start. Always have to backup plan. UPS Inverter should be used during the online meeting so in the case of a power cut, or failure of power meeting will go on smoothly. EZTalks software is useful during an online meeting. Formulate the guidelines of ensuring high-level decorum gets on with it consent smooth running meeting.

Digital Supportive Cultural Environment

Digital work culture plays an important role in any organization. This environment is more supportive of the decision-making process, payroll. The digital environment provides growing and learning environment in any organization. Cloud computing and cloud storage technologies made up of many resources but comprise in one. These technologies are more agility and economical in terms of incurs operating expenses. Cloud services provide a backup facility that is helpful in natural disasters and creating, appending and deleting the file can be done. It is helpful to provide digital security. Cloud storage technologies are more supportive to used importing copying of virtual machines images from the cloud to on-premises location or image library.

Dynamic software available in the market that involves all aspect of collecting, storing, retrieving information in the business or an organization and supporting for Managing the accounts, inventory, payrolls, stocks, sells, banking, financial and other records. For the amendable environment, Management Information Systems related software is specially designed for managing and tracking of any traction. The real-time information accuracy is maintained and tracked by the relational database and for maintain daily schedule and transaction, inventory, accounts, job summary and all records specific software are designed. The pdf file and digital Image processing software, desktop publishing software and enterprises resources planning software and chain management software, etc. are designed for specific purpose, which is supportive of quick, efficient, accurate work and also saves the time. Development speed will be high, easy to use, customer support and ultimately supportive of achieving the goal of the organization.

When you enter data be careful, because once you enter data and click to ok then the automatic output will be shown therefore input data need valid, accurate, and completeness. Install the appropriate software and hardware, should be fulfilling your all requirements (present and future), documentation supporting, error detecting and error handling, reliable. Improper or inappropriate software or hardware creates problem hence, always Install recommended security software with them. Hence, they provide security from hackers and other frauds.

Open-Ended Digital Policy and Approach Release to Resourceful Output for Employees

Digital policies and approached releases to innovative solutions for employees and employer. The organization must be set the clear policies and strategies for the work and their employees with their target relating to the digital environment. The organization must wait while they have to take a decision and become the future policies and approaches that are balancing between the employees and the customer's requirement till releasing the future innovation.

The increasing integration of digital technologies in all aspects of our lives is both a benefit and a challenge for employers and employees.

Security

Digital work culture provides so many facilities for smooth, efficient, qualitative and quantitative work or production with great purpose or growth of the nation. Many of the software is built for greater advancement, and are supportive of achieving the organizational goal in a swift, prompt, easy, more competent with taking less time, give greater and qualitative productivity with efficient, talented employees. Through the man-handling and use for fun or acquire money creates a great challenge for digital work culture. Hence treats and risk like an attack on the person or organization's important data, by the hacking stealing debit or credit card information, damaging person or company reputation or financial devastation, etc. are growing in the cyberspace that creates an importance of the security on digital work culture.

Misuse of the digital signature is handling by the digital certification that proves the ownership of the electronic documents; digital certification means the use of cryptography a set of protocols based on algorithms are used two keys one private key or secret key and another public key. Private or secret key uses symmetric encryption to data transmission that uses the same private key for both encryptions (both sender and receiver are sharing the same key). And Public key asymmetric encryption uses two different mathematical links. It provides email security, able to create a digital signature. Another algorithm has two groups block ciphers and stream ciphers for executing encryption and corresponding decryption.

Ransomware or destruction of information (Brown, 2017) has a great challenge in front of the business. Attackers attack all data stored on disk and convert data into encrypted data. So data will be useless, and then blackmail started after that money demand if the victim gives them money by the anonymity channel they restore the data again. Invaders grab the trade numbers, business plans, social security numbers credit card numbers or password along with creating the fake accounts and create fake transactions to gain money or goods. To handle this problem always create strong and remember you to easy password and install valid anti-virus software on your system.

Malicious software installation, theft or presumption of the password, slowdown or stops together your system regarding problems arises. This problem handled by the purchase of reliable security software. Software must be purchased after reading the review of customers. So that reliability of software will be greater.

Confidential & personal data must secure the encrypted data in the database. Cloud services are also helpful to make secure your business.

A virtual taskforce set up should be responsible for the creating criminal environments, facilitated and targeted by the Internet.

Have a habit of take regular backup, install latest authorized anti-virus software, ignore unknown phone or mobile calls, keep your digital system up to date all can save and secure your digital work culture in your organization and if the company or organization give pay to Information Technology Company for creating safe and secure digital environment of an organization is much better for the tension free organization.

CONCLUSION

Nowadays our human race is going through great revolutionize by digital technologies and devices that modified our lifestyle, values, beliefs and thoughts so-called modern society. Employees can make or maintain family and social contact digitally.

Time has changed, technological developments are growing fast known digital era. Digital media has enlightened the work culture by providing prompt support, comfortable user-friendly and healthy work culture for growing business. Digital Media are supportive for marketing of products or services. Digital media and applications are supportive for decisions making system. Digital decision-making process has been taking place instead of, traditional decision-making process, now. Business angle required complete digital economy. Business promptness and abet can get better by the technology for people to make the ball decision for their unique prerequisite business. Cloud computing solutions are available for machine learning, and automation is beneficial to do daily work and to take or make decisions promptly. Decision-making system required more data-driven software with the advanced security according to the company or organization shortly.

Overall cost-effectiveness for business can be enhanced by ERP and CRM applications and solutions. Whereas trademark TM Applications are used for unique and valid identification of any organization and also help to fill Trademark online.

Digital work culture is more popular in the youth, but they are unhappy and get frustrated when they have to face technical problem regarding digital devices and software. One side, digital work culture has so many positive effects while another side some intelligent professionals with negative mind using digital media and technologies for some wrong things or unlawful work for fulfilling their illicit demands. It creates digital traffic environment and bad culture for any organization. So the bid is that organization, or company head must be blocking some websites for the company security purpose, use technology vs. technology approach for control the digital activities like for electronic data exchange and online communication, etc. can be secured by the insulated cryptographic technologies. Company head/ administration must take the time to time meeting in the light and good environment along with some agenda to create safe and secure qualitative along with the quantitative working environment. Hence organization or company can be acquiring greater productivity. By the appreciating their employee(s) company head / administrator are not only encouraged more and good quality of productivity but also can utilize 3 M's (Money, manpower, and Machines) of the organization along with digital media.

REFERENCES

Brown, L. (2017). *5 Dangers ofRansom ware everyone should know about.* Retrieved from https://www.shredit.com/en-us/blog/securing-your-information/february-2017/5-dangers-of-ransomware-everyone-should-know-about

Duffy, J.E. (2014). Office Suite. *PC Magazine.*

Humbert, M. (2007). *Technology and Workforce: Comparison between the Information Revolution and the Industrial Revolution.* Berkeley: University of California.

Jawadekar, W. S. (2006). *Management Information Systems, Text and cases* (3rd ed.). New Delhi: Tata McGraw-Hill Publishing Company Limited.

Kamath, N. (2012). *Personal Data Privacy in the Online Context. In Computers Internet & E-Commerce, a Guide to cyber laws & The Information Technology Act* (5th ed., p. 314). Delhi: Universal Law Publishing.

Rifkin, J. (2011). *The Third Industrial Revolution.* Palgrave Macmillan.

Sterling, C. H. (2009). *Encyclopedia of Journalism.* Thousand Oaks, CA: Sage Reference. doi:10.4135/9781412972048

KEY TERMS AND DEFINITIONS

Cyber Space: Cyberspace is a virtually created place where users have free to use digital technology along with digital devices for communicating with each other anytime, anywhere, over the globe.

Digital Environment: Digital environment is asocial setting produced through computer technology. This means that people are working and interacting with others virtually by the digital devices and the internet to achieve a specific goal.

Digital Media Resources: Digital media resources are the channels and source of information that is developed for viewing, creating, modifies, distributing and transferring data by the conversion from analog to digital form for communicating with the digital devices (mobiles, smartphones, Computers, tablets, etc.) via computers and network of computers in the way of digital audio, video recording, texting, editing, photography, graphics, broadcasting. On the digital media like electronic magazines, e-books, research e-reports, T.V. news, electronic newspapers, database, blogs, websites, video games, mp3, etc.

Efficiency: Efficiency means capability and competence of employees, machines, system, etc. for effectiveness to produce a qualitative and quantitative output of any organization.

Information: Information is an accurate data or fact that gives meaningful realization that is helpful to achieve the target of any organization.

Office Environment: Office environment means effective background conditions or atmosphere in which a person works and interacts with other persons and a thing for achieving specific objectives.

Organization Ideology and Principles: A systematic vision of belief, thinking, attitude, concepts or values especially about human life or culture, fundamental social cognitions. Defines character of an organization, soul, and spirit of an organization inspires, guide and control the vision of organization where sets of objectives are set by the organization to support for regulating the organization in an effective manner, and also support to planned or arranged work, activities, or events to achieve the goals. Approaches are flows from top to bottom.

Productivity: Productivity in the sense of qualitative and quantitative output from the input of any organization or company provided by the human being as employees, machines, system, etc. depends on their efficiency.

Section 3
Digitalization: Trends and Concerns

Chapter 7
Dynamic Workplace Revolution:
Recent Digitalization Trends in Organizations

Anchal Chhabra
Meera Bai Institute of Technology, India

ABSTRACT

Digitization plays an important role in shaping the workforce and work environment of any organization. It leads to the complete transformation of existing work structure of a particular organization. Digitization can be pertained to different sectors. As the population is increasing and the world is progressing, people are moving fast towards digitization. Today's world comprises of cut throat competition and in order to compete with each other, individuals are consistently working towards invention of such gadgets which will reduce their time so that they can manage their activities effectively.

INTRODUCTION

Digitalization or Computerized change is the utilization of innovation to fundamentally enhance execution or reach of ventures (Bain & Norris, n. d.) – is getting to be a debated issue for organizations over the globe. Officials in all ventures are utilizing computerized advances, for example, investigation, portability, web-based social networking and shrewd installed gadgets – and enhancing their utilization

DOI: 10.4018/978-1-5225-3191-3.ch007

of conventional advancements. For example-ERP to change client connections, interior procedures and esteem suggestions. Different administrators, perceiving how quick advanced innovation disturbed media enterprises in the previous decade, know they have to focus on changes in their ventures. While numerous specialists ask organizations to begin on the advanced change travel, little advice how to do it. This chapter will portray the components of effective computerized change and demonstrate to survey your company's computerized development presently.

The Digital Workplace is the middle of a complex progressed organic group. Whatever the limit in the affiliation furthermore, whatever the business challenges, keeping an eye on the modernized estimation is related to watching out for the Digital Workplace. Changing the standard workplace into a Digital Workplace incorporates reconsidering the working environment, taking control of the dangerous automated progressions, updating capital costs (Chimezie, 1988), while giving another end-customer experience. The Digital Workplace is thus not just an unadulterated IT point; rather it's an open entryway for associations to advance toward getting to be end-customer driven in order to start new execution levers (Digital Transformation, 2017). By doing all things considered the Digital Workplace will be considered as a general association asset, which is fundamental for the business, correspondingly as whatever other focus organizations.

Numerous well-known stories of advanced change include computerized new businesses or, then again cutting edge firms, for example, Amazon, Apple, or Google (Feldman, 1989). Administrators in customary organizations frequently find these illustrations empty. Huge conventional organizations, with many years of history what's more, legacy, are just not the same as these more up to date computerized contestants (Jefferson, n. d.). Other stories concentrate on energizing activities by conventional organizations, yet are narrative and don't demonstrate how those organizations executed the change. These illustrations might be great triggers for different organizations to make a move, yet they give no counsel on how or whether to do as such (Lerer, & Piper, 2003). The working environment always experiences a diversion evolving change, driven by new innovations, around at regular intervals when these advancements – considered "new" for a specific time outline – have at last demonstrated their feasibility to the general population (Lotysz & Helerea, 2014).

Every change is the venturing stone for the following, and missing one makes the climb much more testing. In the previous years, organizations associated their workers together through systems, and in the 2000s, the Web associated these PCs to the world. All through the historical backdrop of advanced coordinated effort, all things considered, associations have been taking after these progressions in their business biological systems either as ahead of schedule adopters or as mindful supporters (Recent Foreign Inventions, 1855). Today, associations are at the edge of a new computerized upset: the time of versatility what's more, the "Computerized

Workplace". A quick business and development move infers that the aptitudes and limits required are an unending moving target (Rosende & Miró, 2013). This is trying relationship to build the capacities they require perilously quick to respond to forceful risks or relentlessly changing customer rehearses. Pay streams and cost structures are up for grabs as cutting-edge change animates (Ryngaert, 2010). Pioneers are presently accepting differing procedures and methodologies for the electronic time. In this present reality, this suggests moving money around, and hypotheses are unmistakably moving towards altering around digital.

Right now, associations are focusing on up skilling the workforce with new capacities and limits, pushing more assignments and activities outside the association through subcontracting and outsourcing electronic limits and methodology (Accenture, 2017). Pioneers are furthermore endeavoring to upgrade perspective of their association as a "propelled supervisor of choice" to attract new capacity. Many gatherings are under-resourced and will subcontract where required. Obviously, the time, effort and cost of retraining a workforce isn't helpful for specific capacities that could be outdated in months. The earth-shattering advancement of sourcing and subcontracting for cutting edge limits and methodology is striking and relentless over the next three years, and makes a more adaptable, dispersed and transient workforce that can adjust to fast cycles of business reevaluation as what's more, when required (Orange Business Services, 2017).

Wherever you look, pioneers in old and new organizations are utilizing innovation in a flood of supercharged advancement that is making mind blowing openings. What's more, it's a business account that stretches out past the standard thing advanced suspects (e.g., Google, Twitter):

- By marshaling machine information around its items, forms and esteem chains, fabricating titan GE now graphs a dynamic future as a product powerhouse.
- Insurance start-up Oscar issues wellness groups to every client and after that use the information to decrease costs for solid conduct. It's not astonishing that clients adore the markdown.
- Banking monster JP Morgan Chase is contributing billions and joining forces with Digital Asset Holdings, a tech start-up, to investigate and enhance he ruin that block chains could wreak on exchanging, advances, installments, contributing and general bank performance.

Flourishing in this computerized time of guarantee and vulnerability implies expanding the speed of advancement, experimentation and coordinated effort. The principle motor for this is not some otherworldly innovation — it's gifted individuals. Drawing in and dealing with the ability expected to make this move will be drastically not quite the same as what worked before. Be that as it may, don't imagine it any other way, innovation will matter — a ton — in light of the fact that new advancements will affect how learning work is composed, circulated and finished. Human ability must adjust to this new reality. The workforce without bounds, hence, must advance in the midst of a computerized insurgency that will without a doubt overturn income streams (SmashFly Blog, 2017). The advanced time will depend on machines; however, winning will require — maybe like never before — ability pools that can flourish in an inexorably digitized economy. Exploring this move will be one of the characterizing achievement criteria for pioneers.

Computerized advancements and what they empower — new plans of action, new income streams and radical new cost structures — are redrawing industry structures and the ability organizations need to flourish (Berge, 2002). Now is the ideal time to get hyper-genuine about how your association's generally vital resource — its kin — will work in this energizing computerized age. A basic for advanced development and engagement has risen: organizations have officially understood that they should utilize computerized channels to draw in with their key partners to keep up significance and drive the discussion (Boorsma, 2017). Notwithstanding, few acknowledge how quick the change needs to happen, or how transformational it should be.

The genuine basic in this present reality where "everything" is digitized is that organizations require seeking after development to upset their own particular plan of action before the opposition does (Corsinovi & Gaeta, 2015). Without development procedures, organizations will lose their upper hand in an undeniably commoditized world. There is no opportunity to lose, as innovation change quickens exponentially and new advanced stages and gadgets are rising. Moreover, the desires of the new 'era Y' or 'advanced locals' imply that organizations must stay aware of the pace of progress or lose importance (Organizational Design, 2017). The test for organizations is to confront the ramifications of computerized change: specifically, the loss of control over the client relationship, expanded rivalry and risk of commoditization, and the need to draw in carefully with providers, accomplices and workers notwithstanding clients.

The beginning stage is an organized way to deal with surveying your advanced development based on a comprehension not simply of the innovation, but rather of the '10,000-foot view' of what computerized engagement intends to the business (Biedrzycki, 2017). To manage this test, organizations must build up a conclusion to an end reaction. They require building up a far reaching computerized procedure and reconsidering their business and working models to convey this.

Importance

Technology refers to the gathering of tools that build it easier to use, create, manage and exchange data. Until the earlier times, the employment of tools by people was for the method of discovery and evolution. Tools remained identical for a protracted time within the earlier a part of the history of human beings however it absolutely was conjointly the complicated human behaviors and tools of this era that fashionable language began as believed by several archeologists. Technology refers the data and utilization of tools, techniques and systems so as to serve a much bigger purpose like finding issues or creating life easier and higher. Its significance on humans is tremendous as a result of technology helps them adapt to the atmosphere. The event of technology as well as laptop technology's web and also the phone has helped conquer communication barriers and bridge the gap between individuals everywhere the globe. Whereas there area unit blessings to constant evolution of technology, their evolution has conjointly seen the rise of its damaging power as apparent within the creation of weapons of all types. During a broader sense, technology affects societies within the development of advanced economies, creating life a lot of convenient to a lot of those who have access to such technology.

1. **Improve Communication:** Business technology is vital as a result of it improves communication within the geographical point. Workplace staff doesn't seem to be restricted to phone calls or inter-office mail to act with each other any longer. Electronic message permits staff to send messages instantly while not interrupting the recipient. Business technology additionally improves communication with purchasers and business partners as a result of data will be responded to multiple channels virtually instantly.
2. **Efficiency:** Office technology saves times by dashing up the work flow method. Digital filing systems save area, paper and printing prices. The utilization of pc systems permits corrections to be created instantly. Resources like electronic files and access to data technology are on the market with the press of a button.
3. **Human Capital:** Technology within the geographic point improves the potency of screening, recruiting and hiring potential candidates. Businesses utilize the net to unfold the word concerning the organization and advertise job openings. Hiring managers will target candidates by victimization digital advertising technology that tracks the websites they visit. Technology like temperament assessments and screening tools permit businesses to work out whether or not a possible candidate is acceptable to the organization.

4. **Mobility:** Technology within the geographical point much eliminates area and time. Video conferencing technology lets businesses on any a part of the world act with each other. Technology reduces travel prices as a result of businesses will get wind of virtual conferences and distribute knowledge while not the necessity to be within the same space. Technology permits businesses to ascertain a world presence at a fraction of the value.
5. **Access to Variety of Learning Resources:** In the era of technology, IT aids lots of resources to reinforce the teaching skills and brainpower. With the assistance of IT currently it's simple to supply audio visual education. The training resources are being widens and widen. Currently with this vivid and huge technique as a part of the IT information, learners are inspired to treat computers as tools to be employed in all aspects of their studies. Specifically, they have to form use of the new multimedia system technologies to speak concepts, describe comes, and order data in their work.
6. **Collaborative Learning:** Now it's created it simple to review also as teach in teams or in clusters. With on-line we are able to be unite along to try and do the required task. Economical communication systems, the phone (fixed and mobile), and numerous recording and playback systems supported technology all have a section to play in instructional broadcasting within the new millennium. The net and its internet sites square measure currently acquainted to several youngsters in developed countries and among instructional elites elsewhere, however it remains of very little significance to terribly more, World Health Organization lack the foremost basic suggests that for subsistence.
7. **Multimedia Approach to Education:** Among the devices used are still and motion photos, filmstrips, television, transparencies, audiotapes, records, teaching machines, computers, and videodiscs. The expansion of audio-visual education has mirrored developments in each technology and learning theory.

Studies within the scientific discipline of learning counsel that the employment of audio-visuals in education has many blessings. All learning is predicated on perception, the method by that the senses gain data from the setting. The upper processes of memory and conception formation cannot occur while not previous perception. Folks will attend to solely a restricted quantity of knowledge at a time; their choice and perception of knowledge is influenced by past experiences. Researchers have found that, different conditions being equal, a lot of data is taken in if it's received at the same time in 2 modalities (vision and hearing, for example) instead of during a single modality. What is more, learning is increased once material is organized which organization is clear to the scholar.

BACKGROUND

Digital Trends in Organizations: Issues, Problems and Controversies

Issues

Digitization at its most straightforward implies the transformation of simple data into advanced data (Hinchcliffe, 2017). As digitization abilities augment, essentially every part of life is caught what's more, put away in a few advanced frames, and we draw nearer towards the organized interconnection of regular items. The effect of this is a constant worldwide trade of data between different associated gadgets.

- **Associations Must Use Propelled Channels to Make Steady Furthermore, Unfaltering Engagement:** With the introduction of the central for the most part used business locales in the 1990s, modernized development passed on another level of convenience to customers (Hinchcliffe, 2017). In any case, automated comfort went to the impairment of critical engagement as cutting edge trades substituted physical affiliations. Besides, automated coordinates make in imprisonment, achieving inconsistency and separation among cutting edge and physical channels. The legacy is one of disappointment for the time being's customers, when for example they find they can't complete an association they began online when they walk around the store. The failure is particularly felt by electronic local people who have grown up used to quickness and convenience likewise, would like to be "known" at every correspondence with an association, paying little personality to the channel in which they secure. Directly, regardless, new advancements exist that make it less requesting than at whatever time in late memory for associations to join channels, and to pass on a consistent engagement indicate whether customers tweet or to take their feet to the street. Imaginative associations are taking the segments of each channel that their customers regard most, and going along with them to pass on a more vital undertaking as a rule (Palmer, 2017).

For example, outline retailer All Saints has joined the solace of web scrutinizing in their physical stores by including web engaged stalls site page, so customers can examine and check openness on-line, and a short time later endeavor the thing on in-store and trick both universes. Apple stores use versatile Point of Sale terminals to incorporate the convenience of 'snap to pay' to their physical shopping foundation and give customers an email receipt from the POS terminal.

- **The Pace of Advancement Change Is Growing Exponentially:** Most associations fathom the need to respond and change in accordance with the propelling usage of development by their customers and other key accomplices (Cognizant, 2017). What they don't comprehend is the methods by which little time they have to address these movements.Advancement determination generally occurs in an S-twist, with exponential improvement of advancement determination from 'early adopters' (Rosa, 2017). Not solely is the scattering of individual headways exponential, however, the rate of scattering is animating, and with each new development release, the 'early adopters' and 'early lion's share' move further to the other side. It assumed control 80 years for telephones to accomplish 60% family entrance, differentiated and 30 years for radio, and 15 for web get to.

Taking after this example, the rate of advancement appointment should continue quickening so that each new advancement outpaces the appointment of its predecessor, and the future will see choice rates measured in weeks and days rather than years. Google+, the new web based systems administration gadget from Google took only 17 days to accomplish 15 million customers, differentiated and 779 days for Twitter and 849 days for Facebook (EY, 2017). It took 15 years for the web to twist up clearly a major and essential bit of consistently life. The future will happen altogether speedier than that.

- **Generation Y:** Expectations and effect Affiliations that have winning as to attracting with generation Y understands that, and moreover stimulating the take-up of new propelled development, this period speaks to additional challenges as a result of their creating wishes (Rick, 2017). Generation Y are best considered not a uniform assembling yet rather as a creating mindset, and the solicitations and yearnings of this social event are floating upwards to more prepared periods, which makes them all the more key to watch. They are today the driving forces besides, the incubation centers of advance, however consistently their effect is creating in customer and expert work environments, suggesting that affiliations must make sense of how to attract feasibly now, or danger being finished off of the beguilement.

To pull in these adolescents, both as buyers and as specialists, an association can't just look incredible – it must be extraordinary. Generation Y requires it all, yet in particular they search for straightforwardness (Rick, 2017). It is no longer possible to express a specific something and be another, as the propelled data trail infers the new period has the instruments and procedures to spread their supposition about a brand virally, and with untold outcomes. This has tremendous repercussions for brand reputation organization and customer relentlessness.

Challenges: Facing the Digital Change Implications

The need to attract painstakingly with suppliers in addition, laborers while associations tend to think about the need to fulfill imparted and cross-channel engagement to their customers, they are less insightful that this need applies also to suppliers additionally, to delegates. The rising of internet systems administration is isolating deterrents between these social occasions. Instead of talking one-on-one to customers, suppliers and agents, affiliations are right now talking with a crowd of people who are often chatting with each other. Also, what's more delivering a considerable measure of uproar, this association infers that information can stream in heading that the association did not speculate (EY, 2017). Generation Y, particularly, have had experiences as customers which affect their wishes in various business dealings, for instance, their collaborations with their supervisors. That infers that, and cost venture assets and productivity benefits, progressed correspondence capacities are transforming into a key weapon in selecting and holding capacity. Gone are the days when a specialist energetically got their new work conveniently through PC and PDA. Today's specialists when in doubt have more information correspondences advancement at their own exchange than they are given at work, and IT divisions are logically saw as containments to their necessities rather than an enabling operator.

Loss of Control Over the Customer Relationship

The increase of mechanized channels and contraptions gives buyers more noticeable access to information, and the strategies for correspondence furthermore, participation. The physical world is being copied in the propelled world through modernized gatherings, associations and assets, on an exceptionally fundamental level changing the way clients interface with associations and each other. Information aggregate and esteem connection have progressed. Buyers now have consistent, adaptable access to data that they as of now relied on upon brands to outfit them with. Today's buyer has steady access to centered evaluating information basically by analyzing a thing institutionalized tag into their mobile phone and examining this information through an application, for instance, Red Laser (Rosa, 2017).

Thus, cost straightforwardness is extending as purchasers can take a gander at the expenses of things moving, setting power unequivocally in their grip. And furthermore, relying upon brands for esteeming information, customers in like manner used to rely on upon brand correspondence and brand qualities to settle on instructed choices, yet this is continuously not the circumstance (Panetta, 2017). Brands are standing up to an incident in the information forcing plan of action and a move towards two-way correspondence and discourse. It is key that brands get a handle progressing in buyer control and find ways to deal with make it work for shared ideal position (Panetta, 2017). This is by virtue of as customers get vitality to pick where and how they interface; they will begin to pick less relationship with which to keep up fundamental associations. This is presumably going to be group arranged affiliations that reflect and interface with their values. Not all affiliations can win the battle to hold the customer relationship – the propelled world has hurled relationship into a completely engaged entertainment.

Extended Contention and the Danger of Commoditization

The companies have successfully recognized the future example of buyers picking to contribute their total acquirement with a tinier game plan of brands. This example is rising in like manner of two forces getting together. Customers require choice, without the cost of making and keeping up various associations; and associations are steadily searching for ways to deal with hold and make extended share of wallet through more significant customer associations (Palmer, 2017). These two forces are gained possible by mechanized ground and globalization. Mechanized channels cut down preventions to section and augmentation globalization, provoking a twisting of expanding contention and commoditization. Innovative affiliations are tolerating the open way to grow, bringing cross-industry blending and clouding of the restricts between endeavors. In advance, physically unmistakable things and divisions now battle with each other, over less unmistakably portrayed customer bases.

Thus, there is an extension of 'super-brands, for instance, Google, Tesco, or Virgin. Starting at now, markets battle with security providers, essentialness association's offer Master-cards and retailers offer essentialness discounts (Hinchcliffe, 2017). Google, is today in direct competition with social sorting out districts, they have also moved into the flexible handset feature, and even the online shape publicize, leaving affiliations in various undertakings contemplating, or perhaps fearing, what Google will do next. The battle to "have" the propelled customer has been hurled totally open.

SOLUTIONS AND RECOMMENDATIONS

Numerous associations are starting at now making responses to the trial of cutting edge by moving from an esteem based to an "interactional" relationship with their customers. They are inviting their customers to wind up some bit of the R&D method, the arrangement handle and the go-to-market. They are moreover consoling them to appreciate the post-bargain reinforce preparation, and misusing of "prosumerism" as an insignificant exertion way to deal with give advantage.

The Prerequisite for New Business and Working Models

There are essential lessons to be picked up from the web agitation (Hinchcliffe, 2017). Firms that have benefitted by its potential results have patched up the models of entire undertakings, for instance, music and advancing, likewise, web engaged arrangements of activity, for instance, dispersed registering have experienced huge improvement. It is this change of business and working models that is anticipated that would stand the trial of time in the propelled period. This lesson is starkly spoken to in the more responsible option by the mistake of various squares besides, mortar arrangements of activity to acclimate to the risk of Amazon (Organizational Design, 2017). To be compelling now, we acknowledge, requires no not as much as the creation of enhanced or new plans of activity, new working models, new strategies for partner with purchasers, and better methodologies for advertising.

A Structured Approach to Manage Transformational Change

Affiliations require a composed approach to manage the trial of partner with key gatherings of spectators in the way that they continuously expect. They ought to grasp the threats and openings that the mechanized world presents to their business. One possible approach is through a propelled improvement assessment, to help expert gatherings to both understand the challenges acted to their particular business and industry and to perceive new, progressed ways to deal with interface with different get-together of individuals. Ernst and Young review propelled advancement using five significant estimations: customers, suppliers and assistants, theorists, laborers and activity (Organizational Design, 2017).

A model like this is not a solution; it is the starting stage on a change travel. It is fundamental to keep up a key separation from over-ensnaring the subject: the fundamental objective should be a sensible appreciation of what modernized infers for the business. With that understanding, it winds up discernibly possible to make a sorted-out movement plan to ensure accomplishment and life expectancy in the mechanized world.

Latest Technologies Used In Different Sectors

Sense and Avoid Drones

While planes and helicopters are needed to be slowed down to be prepared for preparation, a micro drone UAS is place into action directly, with none loss of the precious time.

Their carbon fiber housing permits drones to control with success beneath testing conditions like rain or snow and at extreme temperatures, as within the case of fires. With information transfer to the mdBaseStation going down in real time, it's not necessary for firefighters or alternative support personnel to enter the zone so as to analyze true. The micro drone UAVs will currently fulfill this role. The standard payload system permits varied devices to be hooked up to the micro drone. Employing a gas device, as an example, the smoke composition of a hearth is communicated, so fireplace fighting groups are tuned in to it.

A thermal image camera assists in the search for missing people in forests, fields, etc. and drones are available in an enormous amount of shapes and sizes; however, most come equipped with HD cameras and GPS navigation, permitting them to be flown and half-track remotely. Drones aren't only for fun, though that's perpetually got to be an enormous a part of their attractiveness needless to say. Firms and governments across the globe are becoming interested in using UAVs. Drones, quad copters and octo copters are suited for the needs of aerial inspection. The drones are equipped with HD cameras, heat sensors and transmitters that allow information to be collected. This UAV scrutiny allows common labor which would take days to be eliminated by drones which can even operate in poor atmospheric conditions.

Many drone manufactures have created crafts for nearly every possible scenario and nowadays they're getting used for industrial inspections, aerial photography, agricultural assessments and analysis, 3D parcel mapping and a number of other uses.

IoT (Internet of Things)

The Internet of Things (IoT) has a big result on consumer-facing and industrial enterprises. At the buyer level, this may be seen within the increasing variety of sensor-based good devices flooding the marketplace. Nonetheless the largest economic impact is within the industrial and service-based segments, as well as industrial insurance. By orienting their business needs with the capabilities of the net of Things, insurers will sharpen operational efficiencies, open new revenue streams, drive profitable growth and keep customers shut. Although the construct wasn't named till 1999, the web of Things has been in development for many years.

The primary web appliance, as an example, was a Coke machine at Carnegie Melon University within the early Eighties. The programmers might connect with the machine over the web, check the standing of the machine and verify whether or not or not there would be a chilly drink awaiting them, or they ought to arrange to create the trip all the way down to the machine. Sensible applications of IoT technology can be found in several industries nowadays, together with agriculture, building management, healthcare, energy and transportation. Data entry helps to increase productivity. M2M will facilitate potency around this task and limit mistakes. As a lot of devices become connected, manual entry can fall by the wayside, giving humans longer to concentrate on ways to be proactive with the info instead of time coming into it.

Waiting rooms are ripe areas for M2M interaction: as purchasers wait, cell phones will communicate with in-office devices to provide purchasers with info concerning the approaching appointment, permitting each party to avoid wasting time.

Blockchains

A blockchain is the technology underlying bitcoin and different crypto currencies—it is a shared digital ledger, or a regularly updated list of transactions. This localized ledger keeps a record of every dealing that happens across a completely distributed or peer-to-peer network, either public or personal. Most people use a sure middleman like a bank to create dealing. However, blockchain permits shoppers and suppliers to attach directly, removing the requirement for a 3rd party. Using cryptography to stay exchanges secure, it provides a localized information, or "digital ledger", of transactions that everybody on the network will see. This network is actually a sequence of computers that has to all approve associate exchange before it may be verified and recorded.

A hash is a few things sort of a distinctive digital fingerprint. It is an illustration of a file as a fixed-length string of bits, and an illustration to make sure that nothing causes the destruction of the file. Hashing a block along with the previous block is thus the relevance of a blockchain, which makes it even more durable against tampering with any part of the chain. Ever since the primary Bitcoin dealing was applied in 2009, the digital crypto currency has been a subject of dialogue. While banks and regulators have, for the most part, remained cautious of Bitcoin, the underlying technology of blockchains. The distributed ledger began attracting the eye of banks and startups by the start of 2013. The lure of a block chain was its technique of confirmation and transaction records. Rather than a trustworthy third-party or a financial organization, it depends on an accord among a peer-to-peer network of computers supported by advanced algorithms.

Rather than being hold on during a single information, blocks of time-stamped transactions are held on all systems across a chain. This elimination of middlemen and decentralization of trust has introduced potentialities to create processes like cross-border payments, commerce and settlement quicker, additional reliable and fewer pricey.

Virtual Assistant

A virtual assistant is a freelance contractor which provides personal services to shoppers whereas operative outside of the client's workplace. A virtual assistant generally operates from an office, however is ready to access the required coming up with documents, like shared calendars, remotely. Individuals utilized as virtual assistants generally have many years of helper or workplace management expertise.

Virtual assistants became additional outstanding as businesses increase their use of the net for daily operations. As a result of this a virtual assistant could be a contractor; a business does not have to offer constant edges that it might for a regular worker. Additionally, since the virtual assistant works offsite, there's no want for a table or different space at the company's workplace. (Virtual Assistant, 2011)

"Even while not speech recognition, virtual assistants will give personalized services for people by accessing reams of knowledge. The virtual assistant also can individualize its electronic communication supported by employer's strategy..." (Aquino, 2016). Virtual assistant's area unit is solely as valuable because the knowledge they need access to. And as a lot of devices and systems become connected through the web of Things, a virtual assistant will probably connect those knowledge points to supply extremely personalized services. It conjointly prompt that employee's birthday is springing up and connects with the employee's virtual assistant to collect meal preferences, calendar accessibility, and notice an edifice on Open Table that meets those preferences. It even monitors very important signs which suggest taking a walk when many conferences are taking place by mapping out an optimum route.

Point of Sale

Point of Sale (POS) is very important technological software used in retail sector. It is prominent in many stores, small to medium sized business houses nowadays. POS system can track all of the sales thus it will simply see how the business is doing (at any time). In different words, it replaces the money box.

The overall setup of a POS system includes a server together with one or additional terminals looking on what number checkout stations there are. There's a keyboard for getting into information, a money drawer, credit/debit card swiper, pin pad, and a receipt printer. Several firms conjointly utilize some style of barcode scanner to input data faster. There may additionally be a checkout scale, card reader, and client show. (Mann, 2015).

Some more modern systems supply touch screens for an additional efficient expertise. Mobile and wireless technology could be a fairly-new addition to POS systems and devices like tablets, iPhones, and Androids are getting down to be employed by additional firms. This may be an economical choice for little to mid-sized businesses with restricted wants and may be quite economical. These elements are going to be utilized in conjunction with POS code, which can store data and permit it to be managed inside information. The specifics of code will vary significantly relying upon the trade. Some codes are often used across varied industries, whereas some is niche-centric. POS machine also helps in doing the inventory count like how many units are sold, how many units are remaining etc. Earlier these things were done manually, but now technology in the retail sector has reduced their work, so that they can easily concentrate on other important activities pertaining to their sector.

Many big retail firms like Big Bazaar, Pantaloons, Spar Hypermarket, etc., are using this technology to save more time and improve employee efficiency.

SMART Boards

SMART Boards is a technology used in education sector. It helps in imparting education to students in a smart and effective way. Smart Boards are developed using smart Technologies as associate interactive whiteboard with slightly screen. It had been initially introduced in 1991 as the simplest way for presenters to manage Windows-based applications simply. SMART Boards are finding their means into school rooms, coaching sessions, conferences and board conferences. The surroundings permit instructors and presenters to manage a presentation during an additional versatile means than simply sitting behind a laptop and projector. The projector and Board are connected to the pc. Through the projector, a picture of the pc screen is displayed on the Board. The Board then takes it a step more associated acts as an interactive bit screen monitor for the pc. By touching the Board, the user is in a position to click on buttons, highlight text and drop and drag things right from the Board.

The Board may be connected to the pc either wirelessly or through cables. The wireless affiliation works a similar as Wi-Fi or Bluetooth. Connecting through cables may be through with either a USB or serial ports. Employing a wireless affiliation is most well-liked for larger rooms as a result of it eliminates tripping over twisted cables. The projector is then connected to the pc. The resistive technology employed in Boards is that the same that's used with PDAs. This permits the Board to be able to provide method of the movement of a finger or pen tool once it touches the surface.

The resistive technology is essentially a sheet of skinny resistive on a durable backboard. Once the finger or the pen tool glides over the resistive film, it causes the reflective film to the touch the backboard. This is often then sends as associate analog signal to the pc. In some models, Digital Vision bit is employed for this operate rather than the resistive film, that may be a far better resolution as a result of it doesn't have a similar limitations because the resistive technology.

Cloud Computing

Cloud computing is an important technology which helps in sharing and storing data .Cloud computing could be a method for people and corporations to access digital resources over the net, from regarding anyplace within the world that has property. Cloud computing is usually provided by a 3rd party as a software package service, or is usually designed in-house with DIY techniques and impromptu hardware.

Cloud computing sometimes eliminates or reduces the necessity for on-the-scene hardware and/or software package. As an example, if an individual buys a tough drive backup service that depends on cloud computing, he or she might transfer his or her files through an online association thus they're keeping information on servers that will be settled in another state, or maybe in another country.

Usually the files would be kept in multiple places giving additional security and redundancy that's not possible with normal hardware solutions.

Cloud computing offers the potential to immensely increase on the market resources since some individuals see cloud computing as "IT outsourcing." The thought of outsourcing is especially common within the client industry as a result of firms source their center duties to representatives in alternative places after they aren't ready to notice appropriate client service agents domestically. Cloud computing technology provides a managed service thus anyone can simply specialize in this task, no matter that they are doing that's supported by the service.

The Real Challenges of Digitization Is Not Technology

Clever affiliations have seen that bringing new development into the workplace isn't about hardware or writing computer programs: it's about wetware, generally called individuals. If associations should be the kind of deft business that can profit however much as could be expected from dynamic surges of tech progression, they require individuals who can conform to change (Corsinovi & Gaeta, 2015). The economy is exasperating deliberately at a quick pace and is influencing every limit over the affiliation. Productive corporate automated changes start with specialist getting a handle on advancement. This sends a sensible message to whatever is left of the affiliation that the association is totally serious about executing change (Corsinovi & Gaeta, 2015). In the occasion that modernized change is done well, it starts at the top and moves down all through the association.

Transforming into a propelled affiliation will require new organization capacities combined with interfacing people, methods, data, and things. There is no debating it, transforming into a propelled association is troublesome. Affiliations go up against reams of advance while standing up to digitization (Boorsma, 2017). However, associations that don't move toward that way, and move quickly, they will wind up being a prime plausibility to be disturbed them-self. Exasperate, or be vexed. Every association is an advancement association. If the association don't change, the industry will be changed for the people who still trust that digitization is about making web based systems administration goals or working up a responsive website, here is reality (Berge, 2002): Digital advancements will exasperate regard chains, progressive structures, operational methodology and salary models. Moreover, they will change every industry and every association. Directing propelled change will pick which associations will survive and which not.

Consider how these automated affiliations have vexed endeavors -

Uber: Digitized the taxi understanding and changed the transportation advantage industry.

Airbnb: An absolutely purchaser engaged and propelled encounter plan of activity for friendliness.

Square: Revolutionized convenient portions that could agitate dealing with a record and fiscal organizations.

Netflix: Traditional TV industry has been furious about "over the top" companies like Netflix.

FUTURE RESEARCH DIRECTIONS

Over the span of late years, associations have taken many courses to digitizing the front completions of their associations to make predictable customer collaborations—for instance, building convenient applications that make it less requesting for customers to, say, organize pieces of clothing or open a budgetary adjust. To ensure more vital achievement with those attempts, associations may need to consider ways to deal with digitize their back-office limits, which in numerous associations are dealt with by shared-organizations affiliations. These get-together routinely supervise and pass on particular and administrative support in reaches essential to all claims to fame units in an association, for instance, reserve, HR, and IT. A great part of the time, they undergird focus business limits. In a financial organization setting, for instance, the shared organizations affiliation might be blamed for get ready propel applications or assurance claims. By joining computerization, virtualization, advanced examination, and other propelled progresses into their operations, shared-organizations affiliations may have the ability to streamline shapes. These headways in like manner may engage them to settle on better decisions and improve the way of inside and external customer interchanges (SmashFly Blog, 2017).

Understanding the Open Entryways and Challenges

Most buyers rely on upon some propelled channel to interface with associations. Many say they slant toward shopping through mobile phones or compact workstations for comfort, more conspicuous choice and control, and promising transport of things and organizations, among various points of interest. Subsequently, most associations are controlling new applications, things, and mechanized contraptions that allow them to assemble and separate data, and change bits of learning from those data into definitive cutting edge exercises that can help improve interchanges with customers and business accessories.

Shared-organizations affiliations have a central part to play in this propelled condition. These social occasions exist to streamline the organization of inside methodology and to help the quick and powerful transport of programming and organizations to customers. Dynamically, the back-end work they perform is fundamental for getting data that the parent association can use to make by a long shot prevalent, more predictable front-end experiences.

The common organizations store up at one overall bank, for instance, produced a planned database that gives bargains specialists a sweeping point of view of things and customer slants and activities. Using available customer data and some modernized advances, it could sort sorts of customer activities transversely over item offerings (for instance, credit-and check card application methods and propel application strategies) and present this information in a united view for the parent association, in like manner encouraging trades and improving customer advantage. Greater jumps yield greater increases in the coming year modern associations will keep on taking a hard take a gander at their part in the aggregate esteem chain, regardless of whether they work in a business to buyer or business-to-business plan of action. Makers will focus in on those exercises with the best potential for making esteem, productivity, client reliability, or income.

This vital concentration is generally changing the way of what organizations do, and whom they do it for. Modern manufacturers, Greater hops yield more noteworthy increases In the coming year mechanical affiliations will continue taking a hard look at their part in the total regard chain, paying little mind to whether they work in a business to buyer or business-to-business plan of activity. Creators will concentrate in on those practices with the best potential for making regard, efficiency, customer devotion, or wage. This key focus is on an exceptionally essential level changing the method for what associations do, and whom they do it for. Mechanical makers, for example, are developing their arrangements of activity both upstream and downstream to exploit something past lucrative advantage pools.

Overall, makers are similarly making sections and odds of engagement, for instance, thing personalization, to get closer to the customer. Mechanical wholesalers, in this way, are searching for indispensable accessories to enhance their progressed capacities (e.g., online business, data examination) to counter perils of disintermediation and edge deterioration. For instance, are extending their plans of action both upstream and downstream to take advantage of something beyond lucrative benefit pools. Much of the time, producers are additionally making entries and chances of engagement, for example, item personalization, to get nearer to the purchaser. Mechanical wholesalers, thus, are looking for key accomplices to improve their advanced abilities (e.g., online business, information investigation) to counter dangers of disintermediation and edge disintegration.

Electronic Transformation Delivering Significant Pay Growth, Cost Savings Electronic is the best approach to keeping an eye on the necessities and requirements of present and future on the web clients, who will address half of Asia's masses by 2020, up from 35% today. Realizing this, the pioneers examined for this report are re-adjusting their arrangements of activity to make new things and organizations, bolster pay advancement likewise, and drive new operational efficiencies over their affiliations. Such efficiencies will enable respondents to refocus resources a long way from internal operations and toward improvement openings engaged by mechanized business. The money related impact for associations that go progressed will be tremendous.

Starting at now, outline respondents said their associations have experienced pay advancement of 7.2% as a prompt outcome of their electronic change attempts and they expect that this will more than twofold by 2017. No industry is protected to the impact of cutting edge unsettling influence, based on our investigation revelations. Undoubtedly, even exceedingly controlled organizations, such as cash related organizations, are under extraordinary weight to recast their operations.

CONCLUSION

Another Kind of Purchaser Means Better Methodologies for Doing Business

Today, most associations are either considering or crushing ahead with mechanized change exercises. Every association has a site, and few displaying frameworks are shut down without solidifying electronic long range informal communication. Completely, electronic informal communication is a fundamental some portion of any electronic technique, yet an extensive response to the propelled move must go significantly further. The digitization of everything is a phase change fundamentally more vital than the advancement and gathering of the web, fundamentally in light of its scale and pace of advance. What we portray today as "modernized" in a few years' time will have no necessity for the connecting with word.

A 'propelled camera' is starting at now an insignificant "camera" to the people who know the same. In the same way, a "propelled" framework will move toward turning into old news strategy. This is the reason it is so fundamental to get a head start and learn while there is still time.

Electronics are changing the world, but progress is not immediate. Truly as we are probably aware wireless is no longer just a phone, it is however a potential revolution, we invite relationships to examine what automated pushes mean for them and their accomplices. We reinforce the finish of managers who see electronic essentially more as an opportunity to be tapped than as a danger to get ready for. Not each modernized movement will work for each affiliation, and it is fundamental to assess capacity and cutoff with respect to change some time as of late sending a modernized procedure.

When in doubt, be that as it may, the farther reaching the movement, the more unmistakable the likelihood of accomplishment. We assume that the people who act completely and act now, stay to get the best high ground. The speed of advancement in various endeavors is by and by measured in days or months instead of quarters or years. In these energized conditions, authorities are looking at business sectors, watching contenders and tuning into customers — all with the sound despondency that another tech progression or particularly advanced start-up will shred their arrangement of activity overnight. Using capacity with automated limits is a solution for direct improvement cycles besides, can help give a culture of speed in today's dynamic and precarious business world. Associations that can't proactively hold mechanized instruments, developments and capacity are setting themselves up for a reliable sink to the base, limited moreover saw as unessential by their customers.

Mechanized advances logically choose our normal day by day presence, especially the business world. In this regard, authorities agree that IT can give useful high grounds that basically affect corporate accomplishment. While the importance of innovative advances reliably extended, IT frameworks were generally treated as subordinate to business frameworks: Practitioners as well as experts called for business-IT course of action, which emphasizes the business estimation of IT furthermore its some portion of supporting business framework. Lately, the thought of (DBS) went to the fore, proposing a merger of business and IT procedures as a basic for driving headways and remaining centered. This ponder constitutes an overall paradigmatic move in understanding imperative organization in the season of automated money related matters, speaking to the change of things, organizations, shapes, legitimate structures and furthermore business models through imaginative advances.

Since business and IT methodology should no longer solely "supplement" each other, it is fundamental to assess how the blend of business and IT impacts affiliations and their strategizing shapes.

REFERENCES

Accenture. (2017). *Technology Trends 2017 – Accenture Tech Vision*. Retrieved March 31, 2017 from https://www.accenture.com/us-en/insight-disruptive-technology-trends-2017

Aquino, J. (2016, August 24). Artificial Intelligence and the Future of Work. Retrieved July 21, 2017 from http://www.1to1media.com/technology/artificial-intelligence-and-future-work

Bain, L., & Norris, S. (n. d.). Assessing the impact of applications of digital health records on Alzheimer's disease research.

Berge, Z. (2002). Obstacles to distance training and education in corporate organizations. *Journal of Workplace Learning*, *14*(5), 182–189. doi:10.1108/13665620210433873

Biedrzycki, N. (2017). *Digitization is not a trend: Digitization is already our reality*. Retrieved March 31, 2017 from https://norbertbiedrzycki.pl/en/digital-journey

Boorsma, B. (2017). *Digitization: Challenges & Opportunities. blogs@Cisco-Cisco Blogs*. Retrieved March 31, 2017 from http://blogs.cisco.com/government/digitization-challenges-opportunities

Chimezie, A. (1988). The challenge of responsibilities: Confronting the revolution in workplace rights in modern organizations. *Employee Responsibilities and Rights Journal, 1*(1), 5-23. doi:10.1007/bf01385449

Cognizant. (2017). *People — Not Just Machines — Will Power Digital Innovation.* Retrieved March 31, 2017 from https://www.cognizant.com/whitepapers/People-Not-Just-Machines-Will-Power-Digital-Innovation-codex1850.pdf

Corsinovi, P., & Gaeta, D. (2015). *Managing the Quality Wines beyond Policies and Business Strategies. Review Of Contemporary Business Research, 4(1).* doi:10.15640/rcbr.v4n1a3

Digital Transformation. (2017). *A Roadmap For Billion-Dollar Organizations.* Retrieved March 31, 2017 from https://www.capgemini.com/resource-file-access/resource/pdf/Digital_Transformation__A_Road-Map_for_Billion-Dollar_Organizations.pdf

EY. (2017). *The digitization of everything How organizations must adapt to changing consumer behavior.* Retrieved 31 March 2017, from http://www.ey.com/Publication/vwLUAssets/The_digitisation_of_everything_-_How_organisations_must_adapt_to_changing_consumer_behaviour/$FILE/EY_Digitisation_of_everything.pdf

Feldman, D. (1989). Careers in Organizations: Recent Trends and Future Directions. *Journal of Management, 15*(2), 135–156. doi:10.1177/014920638901500202

Hinchcliffe, D. (2017). The new digital workplace: How enterprises are preparing for the future of work. *ZDNet.* Retrieved March 31, 2017 from http://www.zdnet.com/article/the-new-digital-workplace-how-enterprises-are-preparing-for-the-future-of-work

Investopedia Staff. (2011, January 19). Virtual Assistant. Retrieved July 21, 2017 from http://www.investopedia.com/terms/v/virtual-assistant.asp

Jefferson, O. (n. d.). Workplace Equality in International Organizations: A Way Forward. *SSRN Electronic Journal.* doi:10.2139/ssrn.1236602

Lerer, L., & Piper, M. (2003). *Digital strategies in the pharmaceutical industry.* Houndmills, Basingstoke, Hampshire: Palgrave Macmillan. doi:10.1057/9780230598799

Lotysz, S., & Helerea, E. (2014). Programme and abstracts: technology in times of transition. In *Proceedings of the 41st ICOHTEC symposium,* Transilvania University of Brasov, Brasov, Rumanía.

Mann, N. (2015, October 26). How Does a POS System Work? Retrieved July 21, 2017 from http://www.businessbee.com/resources/profitability/pos-system-work

Orange Business Services. (2017). The Next Workplace Revolution– how technology is changing jobs. Retrieved March 31, 2017 from http://www.orange-business.com/en/blogs/connecting-technology/innovation/the-next-workplace-revolution-how-technology-is-changing-jobs

Organizational Design. (2017). *Digital Technologies and Trends 2017-How Organizations Manage Innovation-Organizational Design.* Retrieved 31 March 2017, from https://gernotkapteina.com/2017/01/17/digital-technologies-and-trends-how-organizations-become-digital

Palmer, M. (2017). *8 Innovation Trends for Digital Insurance in 2016. The TIBCO Blog.* Retrieved March 31, 2017 from http://www.tibco.com/blog/2016/01/15/8-innovation-trends-for-digital-insurance-in-2016

Panetta, K. (2017). *Gartner's Top 10 Strategic Technology Trends for 2017-Smarter With Gartner.* Retrieved March 31, 2017 from http://www.gartner.com/smarterwithgartner/gartners-top-10-technology-trends-2017

Scientific American. (1855). Recent Foreign Inventions. *Scientific American, 11*(1), 3-3. doi:10.1038/scientificamerican09151855-3a

Rick, T. (2017). *The real challenge of digitization is not technology.* Retrieved March 31, 2017 from https://www.torbenrick.eu/blog/technology/the-real-challenges-of-digitization-is-not-technology

Rosa, F. (2017). What do digital; automation and outsourcing have in common. *CIO.* Retrieved March 31, 2017 from http://www.cio.com/article/3020058/leadership-management/what-do-digital-automation-and-outsourcing-have-in-common.html

Rosende, M., & Miró, M. (2013). Recent trends in automatic dynamic leaching tests for assessing bioaccessible forms of trace elements in solid substrates. *Analytical Chemistry, 45, 67-78.* doi:10.1016/j.trac.2012.12.016

Ryngaert, C. (2010). *The Immunity of International Organizations Before Domestic Courts: Recent trends. International Organizations Law Review, 7(1), 121-148.* doi:10.1163/157237310x523803

SmashFly Blog. (2017). *3 Trends Defining Hiring Success In The Digital Economy-SmashFly Blog.* Retrieved March 31, 2017 from http://blog.smashfly.com/2016/04/14/3-trends-defining-hiring-success-in-the-digital-economy

KEY TERMS AND DEFINITIONS

Android: Android (stylized as android) is a mobile operating system developed by Google, based on the Linux kernel and designed primarily for touch screen mobile.

Bit Coin: Bit coin is a crypto currency and a payment system invented by an unidentified programmer, or group of programmers, under the name of Satoshi Nakamoto. Bit coin was introduced on 31 October 2008. The idea was to produce a currency independent of any central authority, transferable electronically, more or less instantly, with very low transaction fees which was an electronic payment system based on mathematical proof.

Cloud Computing: Cloud computing is a term referring to the delivery of hosted services over the internet. Public, private and hybrid are the main cloud computing.

Cloud X: X, an American semi-secret research-and-development facility founded by Google in January 2010 as Google X, operates as a subsidiary of Alphabet Inc. X has its headquarters about a half mile from Google's corporate headquarters, the Googleplex, in Mountain View, California.

Digital Vision: A design of a technical architecture which will meet the needs of the business vision.

Fit Bit: Fit bit (NYSE: FIT) is an American company headquartered in San Francisco, California, known for its products of the same name, which are activity trackers, wireless-enabled wearable technology devices that measure data such as the number of steps walked, heart rate, quality of sleep, steps climbed, and other personal.

I Phone: I Phone is a smart phone made by Apple that combines an iPod, a tablet PC, a digital camera and a cellular phone. The device includes Internet browsing and networking capabilities.

Multimedia: Multimedia is content that uses a combination of different content forms such as text, audio, images, animations, video and interactive content.

Tablet: A tablet is a wireless, portable personal computer with a touch screen interface. The tablet form factor is typically smaller than a notebook computer, but larger than a smart phone.

Touch Screen: A touch screen is a display device that allows the user to interact with a computer by using their finger. Some touch screens use a grid of infrared beams to sense the presence of a finger instead of utilizing touch-sensitive input.

Chapter 8
Uber Strategies for Transport Incorporated Private Market

José G. Vargas-Hernández
University of Guadalajara, Mexico

L. E. Fabián Rojas Álvarez
University of Guadalajara, Mexico

ABSTRACT

The aim of this paper is to analyze the different strategies that take Uber to join the global market successfully, positioning itself in different countries and analyze how come these businesses and strategies that follow to become successful to the extent that Uber is doing, not just one city but in several countries around the world. In order to accomplish this, it is necessary from a previous literature review that is appropriate to identify the different theories that may be applicable, taking terms as work global, you tried to see Uber as a technology - based company and see it from an overall, same strategy refers to a strategy that follows the company having a worldwide standardized product, another issue that would revise the importance of the theory of institutions.

INTRODUCTION

For decades, the transfer of people from one place to another in large cities has been using public transport, taxis, or those with better economic position and resource allows them to move in their own car. That was the traditional way going from one point to another, but with the different changes that occur in society today, technological, economic or related to the environment issues have been marking

DOI: 10.4018/978-1-5225-3191-3.ch008

new ways of moving from one point to other issues. These new forms also have to do with issues of passenger safety, comfort and other factors.

As a result of these new ways of moving emerges UBER, which for several years start to mark a new trend in moving people from one point to another, especially in big cities around the world. It has its origins in a revised class concept of Strategic Management as it is born global that refers to those companies that are born in a global or international markets and are marking a new trend in the way of doing business, taking advantage of the technological changes that have occurred in the world in which today we live. This company considers having a relationship with the technological-based firms that are characterized by being innovative, enough reason to learn more about the same as it was able to study from approaches such as the theory based on institutions and global strategy.

Uber has its origins in 2008 as it tells the story in its page when Travis Kalanick and Garrett Camp could not get a taxi in Paris, which led to an idea to create an application where with just one button, could have a trip. https://www.uber.com/our-story/ What started with this simple idea is now changing the way people are transported in large cities? According to the Uber definition is the smartest way to move and luxury at low cost.

BACKGROUND OF THE PROBLEM

Today globalization plays an important process, because by this concept can be eliminated some barriers between countries. Through technology has been a way these barriers have been eliminated, since technological innovations often come equally to different countries and at different times around the world, so it can be seen different products with similar characteristics, which is called as a standard product, regardless of whether they are produced in different cultures.

The story so before it started this century tells that the common way to establish a company along the story is that the company or the plant reached the place where he would have operations, had a physical place, and was established formally. Now it is seeing a new trend in strategies and ways of doing business in enterprises as a platform generated by its operations, profits, sales and everything related to business dedicated activity. Which have started with this simple idea is now changing the way people are transported in large cities, according to the Uber definition is the smartest way to move and luxury at low cost.

Technology and innovation play an important role in the process of globalization. Adopting a comprehensive strategy for technology companies is essential and addressing new ways of doing business. Nowadays have been emerging these new companies of successful technology-based, born in markets international companies like Netshoes, Netflix, Google, Amazon and many other technological changes that have been created and have managed to be successful in their business segment. The creation of these new companies has diversified options that consumers have to purchase the good or service that require, with new technology and not in the common traditional form.

One problem faced by companies that wanted to provide the services of private transport in the country, were the unions of taxi drivers that prevent or put barriers to companies that wanted to provide these services faced, and who would provide the service had to join the unions of transportation in addition to the permits they had to acquire by the authorities to provide the service. Thus, the institutions were an impediment to a company that wanted to be in this area. But also, an advantage of same weaknesses they could have the same institutions, they have created companies like Uber and reaching nations that did not consider these businesses in their regulations. Taking advantage of those holes is how they have succeeded in entering the global market.

Within technology and new ways of doing business in the globalization process, it can be found different definitions and ways in which this process is changing. One of these definitions is that of Friedman (1970) and for example he says that globalization is a phenomenon characterized by technological innovation and its appearance in the world has been consolidated in stages. The initial stage of economic globalization, for some authors, is the industrial revolution the event that gave rise to this phenomenon. Most authors state that consolidation was given to a process of trade integration (Carbaugh, 2005).

In recent years this type of technology-based companies has increased their presence around the world and have led to the emergence of more of this type and have seen more and more not only born but become successful, not only in a nation but in the global market place.

DELIMITATION OF THE PROBLEM

A technology-based company can differentiate itself from other companies with the main feature innovation processes, either in the product or service offered. They base their activity on technological research, having scientific and technological knowledge.

For purposes of this research, the case of Uber is studied from the side of the strategies used to enter the global market, and succeeding in many countries, taking into account that each country and culture is different, the use of concepts as work global, technology-based companies, global strategy was considered a factor for growth and establishment of industries or companies of this type.

The technology-based companies like Uber have achieved their positioning in the limited control of institutions to such companies, in the absence of regulatory schemes both Uber and other companies have used this to establish their operations. Some features that make them different companies like Uber is that they have fewer staff operating or working for the company, since by creating the online platform and working and operating costs are lower than in a company established in a physical place also presented in the world with a standard or homogenized product.

Companies like Uber generate an advantage for users or consumers and have achieved different countries or states to consider making changes to their laws to fit the modus operandi of these companies. States allow or block entry of Uber and other companies in their type, having a power in the region where they arrive with their operations, thus having influence on decisions or regulations made in the laws for operation.

The research of these report wants to answer is on How technology-based companies with a global strategy have succeeded in entering and being successful in the new way of doing business?

The importance comes to have this work of serious research to understand the impact of the strategies of technology-based companies taking specific case of Uber. This is done using the theory of institutions and by applying the concept of global strategy came to succeed. Overall, a case of Uber especially as famous as the raised here in this paper case where the company has already been established in at least 20 countries, albeit with some adjustments as the nation where the product offering remains standard worldwide reaches, and despite having problems in some cities where he settled.

These problems especially with local competitors, called taxis are mainly because they think is unfair competition and with innovation that the company Uber made in the product achievement that many people with common form move from side to side in public or private transport, now do differently and have changed the way they use the shuttle.

While we know that the group of people or consumers who can access this service is limited, because people who use it so far require at least two basic things to make use of the product. One of them is to have a cell phone where downloading the application, then also required to have a bank account as a credit card or debit card with which to pay the service they have. These basic elements are essential of this new business model and so far fundamental to acquire a service like what Uber offers, because without them they cannot access the transfer from one place to another. Thus, this is being a limiting factor that not all people can consume this good or service, so the service is intended for a specific sector of the population which operates the platform.

JUSTIFICATION

The technology-based companies today play an important role in the business world, which tries to capture this document as through its incorporation and importance have implemented a new way of doing business in companies of its kind, reducing costs, applying innovation in its services and creating a new business model, and in some cases, taking advantage of the weaknesses or gaps that exist in the laws and institutions of the nations where they arrive. Companies like Uber were not considered by institutions or legislation. Thus, taking advantage of the gaps, these companies have managed to enter the market and in some cases adapted in certain respects, but retaining the service innovation.

In the case of Mexico, so far is not entirely clear the legal fulfillments that must have this company. Mexico is working on making changes to laws, regulations roads and adaptations of the legal framework of the relevant institutions to consider Uber and companies of its kind in formal enterprises. In this case, some state laws are working to make adaptations to their traffic laws, such as the state of Jalisco and Mexico City that having already advanced and concrete proposals as to how to resolve this issue. However, at the time of writing this report, some of the proposals were being approved or published to make them official.

Moreover, companies offering the same service as Uber, in this case the taxis see it as an illegal and unfair to the service competition that for years have been offering and operating in some form as a monopoly. They do not pay taxes and have adopted certain aggressive strategies to try to reach the market to Uber through intimidation or aggression towards users or drivers of the business. Taking advantage of technological changes and the weakness of the institutions not considered. Also with the emergence years ago of other technology companies and carrying out a global strategy have been incorporated to market, different companies.

One of these technology-based companies with a comprehensive global strategy that has carried out these concepts in a successfully way is Uber because today it is seen with these strategies that has-been established in the market. After seeing this case have emerged more companies with the same features and offer the service of private transport, as are Cabify and City drive to name some of the same sector. Even its operations are similar to the case analyzed in this work. These firms are technology-based companies that offer a private shuttle service to their users through a platform. Although not all companies operate as such in the same cities, their strategies are like.

On the other hand, if it is seen the part where it operates each company taking the case of Mexico. For example, in Guadalajara operates Uber and City drive but not Cabify, leading to the conclusion that even with global strategies are not present in the same cities or arrive at the same time. Uber is in Guadalajara and 379 cities worldwide and is aimed at a specific segment of the population that is people with bank account, credit card or debit card to pay for the service they purchase, as well as having the application installed on their cell phone.

Within its business model, Uber does not accept payment in cash so far, this being an important innovation in the model, and this aspect is also part of the diversification of the service. Another important issue is quality standards for these keep the user since the end of each trip it can rate the service and the driver, this being another innovation in the service provided how had offered traditional way.

An important issue in such companies as already mentioned is the diversification of product or service offered, as it has been key to succeed and companies like Netshoes, Netflix, Amazon and the company analyzed in this case Uber. These technology-based companies have in common is that they are based on diversification and innovation in what they offer which is considered the most important to have successful operations in a globalized world.

An important segment is the population that has access to the good or service that is launched. when somebody wants to buy a product like shoes, food, books or any other product only they go to the market place where the product is sold and with money in hand it is acquired and anyone who can pay the price can buy it.

In the case of service Uber, this simply fact does not apply because it is a service that only is acquired it from a strict sense by people over 18, who are the people likely to have a bank account to pay for the service. Thus, the target population or that can be client f Uber, are people over 18 years with bank account. Uber on its website states that offers the service in five municipalities of the Metropolitan area of Guadalajara (ZMG), that is Guadalajara, Zapopan, Tlaquepaque, Tonala and Tlajomulco, taking a total of 2,745,260 potential customers broken down as follows the municipalities, and displaying data as much as by men and women.

According to data from the Metropolitan Transportation Institute Jalisco, taxis continue to maintain market power and Uber is positioned in second place in service of private transport as it is concerned. An advantage of Uber, considered by users is mainly the price and greater security.

THEORETICAL ASSUMPTION

One aspect that was taken into account in writing this paper, it was to consider the different theories and revised concepts. One of the aspects that fit for this work was the subject where the importance of institutions is played and whether it can reach

Table 1. Total population of the metropolitan area of Guadalajara.

Municipality	Population 18 and over	Masculine population 18 and more	Feminine Population 18 and more
Guadalajara	1,031,902	482,713	549,189
Tlaquepaque	376,022	182,406	193,616
Tlajomulco de Zúñiga	235,703	115,015	120,688
Tonalá	289,679	147,418	142,261
Zapopan	811,954	388,501	423,453

Source: Prepared with data from INEGI 2010 census.

affect or influence strategies that take companies to conduct their operations. An aspect that Uber considered is the adaptation to the current legal framework, which call as the theory of institutions, and in this category, it is found a definition of the Nobel laureate in economics, North (1990) who defined an institution as humanly planned restrictions that structure the interaction of people, which is popularly known as rules of the game.

Likewise, the same author classifies 2 different types of institutions and classifies them as follows (North, 2005) formal and informal institutions. In the former are the laws, regulations and standards, and on the side of informal talks about rules, culture and ethics, including different ideas, values and attitudes of people in their behavior in society. Moreover, it is possible to distinguish three aspects the authors belonging to the first generation of institutionalisms, neo institutionalisms and the new institutional economics (Urbado and Hernandez, 2007). Moreover, Veblen (1965) also talks about the institutions and gives us another definition that by the end of this work is also acceptable. It is defined as common and predictable patterns of behavior in society, including the habits of thought and action generally shared.

One area where have failed Institutions are flaws in the regulations for these type of companies, i.e. not governments had considered the new e-business, category where Uber is characterized. Canals (1994) states that the objective of internationalizing a company are the opening of new markets, lower production costs, and a more efficient structure of production and distribution of the company.

The business model of this type of business is focused on a specific sector of the population and from this part gives a new way to operate, which is not available for all people who want to use the product or service this type of companies offer. The overall and global strategy adopted by these firms have advantages over competitors as it refers to how to compete, and focuses on providing standardized products and services worldwide (Peng, 2000).

Within the theory of institutions see different regulations that have occurred around the world to allow the application to offer their services. For example, in the United States in the city of Chicago a new category of transportation calls service transport networks, just as in Mexico City was created have made regulations for the operation of the service, in New Orleans you will have to pay the city by permits, something similar happened in New York where the council of citizenship and Uber reached an agreement to set the number of vehicles that can circulate in the city.

Just as there are cities where they operate without any regulation of institutions, there are cases like the above where they operate with certain regulations by the institutions and authorities of the city, but is also the case of Florida in the United States because it does not operate in this city for failing to meet the requirements of city tax.

On the other hand, the phenomenon of globalization (Peng, 2012) refers to the close integration of people and countries around the world, and today this concept is applied in different sectors and industries.

Within the theory of institutions are seen different regulations that have occurred around the world to allow the application to offer their services. For example, in the United States in the city of Chicago was created a new category of transportation called service transport networks, just as in Mexico City was created, both have made regulations for the operation of the service. In New Orleans, the customer has to pay the city permits. Something similar happened in New York where the council of citizenship and Uber reached an agreement to set the number of vehicles that can circulate in the city.

Just as there are cities where they operate without any regulation of institutions, there are cases like the above where they operate with certain regulations by the institutions and authorities of the city. But, it is also the case of Florida in the United States, because it does not operate in this city for failing to meet the requirements of city tax.

On the other hand, the phenomenon of globalization (Peng 2012) refers to the close integration of people and countries around the world, and today this concept is applied in different sectors and industries.

COLLABORATIVE ECONOMIES BUSINESS MODEL

Bostman & Rogers (2010) (2010) argue that firms like UBER are part of a classic model of collaborative economy, being a disruptive innovation (Christensen & Raynor, 2003) that occurs when individuals share common interests and a common philosophy of life. In this case, they aim to rent cars in shared taxis or whole taxis through a social search and management system (Bostman & Rogers, 2010). The emergence of this phenomenon is possible as the evolution of technology allows imitate the exchanges that usually or used to give face to face, on a larger scale thanks to the internet, together with the ability to create trust between strangers.

Rifkin (2014) points out that the decline in marginal costs, which tends to be near zero, is resulting in a dichotomous economy, partly capitalist market and on the other hand, collaborative commons gradually drawn out a new economic paradigm. Interian, (2016) argues that the sharing economy is credited with reducing transaction costs, increase efficiency and promote accountability and competence. This model does not require a centralized entity that should have an inventory and therefore is free of logistics, costs associated with maintaining inventory, product and geographic expansion for these purposes.

This model by allowing individuals to take advantage of the ability to take advantage of an asset that already possess, collaborative business model eliminates in an efficient and convenient way transaction costs. The specific characteristics of collaborative consumer of passengers and commuters are little known, although generally, it may try to travelers connected and experts, likely to responsible consumption, characterized by a high level of trust in the other members of the community and familiarity with internet and online commerce, demanding about the quality of its transport and commuting experiences (Russo & Quaglieri, 2014).

Despite its purported benefits, companies of collaborative economies have been strongly criticized for the way they operate, and in some cases, have been forced to shut down operations. Critics argue that the laws are evading created precisely to regulate certain practices in which their companies are engaged; users sharing platforms become "prosumers" people who consume, like produce (Streitfeld, 2014)

Blurring the line between consumer and producer collaborative economy breaks with the traditional business model that companies own and people consume. On the other hand, the concept of sharing erodes the disinterested public regulation, substituting private regulation or leaving unattended regulated transactions. Because of this, business creation becomes more widespread and may even lead to displace their regulated and established counterparts some time ago. An example of this is Uber, which has become a ubiquitous service in major cities around the world (Interian, 2016). The most important cities in the United States and Europe have begun to implement regulations regarding share or rent cars and have initiated investigations with the goal of bringing the collaborative economy in accordance with existing laws (Chafkin, 2016).

On the other hand, information technologies have facilitated the exchange of user experiences, enabling the comments and product valuation objective and transparent manner. These valuations are perceived in the market as certain and are changing consumer behavior and redefining the role of influence during the buying process by providing more realistic expectations (Cañigueral, 2015). That is when the consumer can know the costs and perceived by others who have commented and valued their consumer experiences, which later will help to make a decision with some confidence, even if the consumer never had before contracted this service, reducing significantly benefits the perceived risk (Wen, 2009).

TECHNOLOGY-BASED COMPANIES, THEIR BEHAVIOR AND MARKET PERFORMANCE

Today, economies have adapted to new business models or at least they try in some cases that have been created with this type of technology-based companies. For the same, now no longer have the need to have a physical space, plant or building in which to carry out its processes, but the base of their operations is through a platform in the network that can be accessed from any device with internet, what we call a new model of e-commerce business.

In its business model (Uber 2016) charges a 20% or 25% in the case the last drivers to register in the system to make use of the platform and the remaining is for the driver. According to the company in the metropolitan area of Guadalajara there are about 3,000 registered vehicles.

The success of Uber in the market is due to several factors. It is clear that neither it does not think and created anything new, nor does not offer a service that did not existed, or invented anything. But what it did was to apply an innovation to a service that already existed, introduced a new way of doing business and using technology achievement join and establish successfully its operations with a standard product around the world, keeping costs down and driven by different factors that in the analysis of results will be more clear.

ANALYSIS OF RESULTS

Today many companies seek more customer satisfaction or give greater importance to this issue. Therefore, they carry out different strategies to accomplish this and satisfy their customers. According to the sector where companies unfold will make the appropriate innovations to the product or service offered by adaptations according to geographical area, with strategic advertising, by price or by improving product quality. Sometimes what the customer is looking for comfort, closeness or facility to acquire the asset or service. This is what UBER has done to offer its service. Uber implements innovations in the business model.

In finding that customer satisfaction, Uber relies to implement its innovations both in service and customer experience that maximizes utility when using the service. One of those innovations was the automatic payment. Also requesting service by an application installed on the user's phone that easily and at the touch of a button is the means to obtain the service. So, consumers identify this innovation in service and thus the company could exercise monopoly power over its competitors once started. Uber to operate the platform seized market share by displacing the service that for years has been given in different parts of the world called taxis that were intended to transport people who had no car or prefer not to use it.

With all the technological changes that have occurred in recent years, it is common that most of the population have access to a cell where through download the application and use the platform can request their transportation service, regardless of whether they bring cash. Uber main advantages over its competitors and by its strategies just identified the customer, are:

1. Innovation in the way of providing the service.
2. No need cash (automatic payment).
3. The customer requires a cell with the application and a bank account where he will be charged by the service.
4. Cheaper than a taxi fare.
5. Provide a standard or homogenized product.
6. They offer a bonus in addition to transportation, like listening to the music the customer wants during the trip, water, air conditioning in the car.

An important aspect to consider is that many innovations and changes that a company has made if the service or product offered does not meet the quality standards of its customers would not be accepted. Bringing the theoretical aspects to the practice when technological innovations, prices and product quality were bad, Uber would not have achieved success that has today. In this way, it can be seen how by implementing a comprehensive and global strategy, a standardized product and application of technology in its service, Uber achieved to implement a new business model in the field of private transport service, and how this has given rise to emerging technology-based companies in different industries or sectors.

These technology-based companies begin to take an important role by having greater technological resources in their operations, with so many changes that have occurred in the way business enterprises have had to adapt to these technological changes and because of this have emerged companies like Uber.

CONCLUSION

One of the advantages of technology-based companies like Uber is to reduce costs and adapt to the new way they are giving businesses through different mechanisms and strategies that were studied for this case. It can be concluded that through a global strategy and technological adaptations companies like Uber have been successful and also taking advantage of the gaps that institutions not considering such firms have adapted to the country coming through a standard product offering in the world.

Companies of its kind have well defined characteristics as that base their operations through the use of technological resources, maintains a standard or homogenized product regardless of the region or geographical area and implements global strategies. However, a weakness that can have companies in their type and it is necessary to assess the institutional aspects, as in some countries are making changes to allow operation through restrictions or in some cases block and not allow its operation. It is important to consider carrying out a successful global strategy.

Another advantage that can have the technology-based companies over their competitors is that their operations are based on the use of technological resources, so it can be concluded that through technological innovations and through a global strategy, it is shown that there is an emerging new type of economic system or a new business model that emerged a few years ago and through successful cases like Uber. This new model begins to change the course of business.

The elements that the company included and considered fundamental are the implementation of automatic payment, request the service by installed on a cell application because without it cannot access the transfer from one place to another, thus being a limiting so that not all people have free access to the service.

Finally, it should be mentioned that innovations do not reach at the same time to all places. Where the technology-based company starts operations, a company is critical to have a successful experience away from and continue to apply the strategies used to bequeath to success in its home and then expand their horizons.

REFERENCES

Bostman, R., & Rogers, R. (2010). *What's mine is yours: the rise of collaborative consumption*. New York: Harper Business.

Canals, J. (1994). *La internacionalización de la empresa*. Madrid: McGraw-Hill.

Cañigueral, A. (2015). Consumo colaborativo. *Leaners Magazine*, *2015*, 18-22.

Carbaugh, R. J. (2005). La globalización de la actividad económica. In R. J. Carbaugh (Ed.), Economía Internacional (pp. 4-8). Thompson.

Chafkin, M. (2016, January). Can AirBnB unite the world? *Fast Company*. Retrieved from http://www.fastcompany.com/3054873/can-airbnb-unite-the-world

Christensen, C., & Raynor, M. (2003). *The innovator's solution: creating and sustaining successful growth*. Boston: Harvard Business Press.

Friedman, M. (1970, September 13). The Social Responsibility of Business is to increase profits. *The New York Times Magazine*.

INEGI. (2010). Mexico en cifras. Retrieved from http://www.inegi.org.mx/Censos(2010)

Interian, J. (2016). Up in the air: Harmonizing economy through airbnb regulations. *Boston College International and Comparative Law Review*, 39, 129–161.

North, D. C. (1990). *Institutions, institutional Change, and economic performance*. New York: Norton. doi:10.1017/CBO9780511808678

North, D. C. (2005). *Understanding the Process of Economic Change*. Princeton, University Press. doi:10.1515/9781400829484

Peng, M. W. (2000). *Business strategies in transition Economies* (pp. 42–44). CA: Sage. Pp.

Peng, M. W. (2012). *Estrategia global*. México: Cengage Learning.

Rifkin, J. (2014). *La sociedad de coste marginal cero: el internet de las cosas, el procomún colaborativo y el eclipse del capitalismo*. Barcelona: Paidós.

Russo, A., & Quaglieri, A. (2014). La lógica espacial del intercambio de casas: una aproximación a las nuevas geografías de lo cotidiano en el turismo contemporáneo. *Revista electrónica de georgrafía y ciencias sociales*.

Streitfeld, D. (2014, October 15). Airbnb listing mostly illegal. *New York Times*.

Uber. (2016). Retrieved from https://www.uber.com

Urbado, D., Díaz, J., & Hernández, R. (2007). Evolución y principios de la teoría económica institucional. Una propuesta de aplicación para el análisis de los factores condicionantes de la creación de empresas. *Investigaciones Europeas de Dirección y Economía de la Empresa*, *13*(3), 183–198.

Veblen, T. (1965). *La teoría der la empresa de negocios (1904)*. Buenos Aires: Eudeba.

Villarreal, O. (2005). La internacionalización de la empresa y la empresa multinacional: Una revisión conceptual contemporánea. *Cuadernos de Gestión*, *5*(2), 55–73.

Wen, I. (2009). Factors affecting the online travel buying decision: a review. *International Journal of Contemporary Hospitality Management*, *21*(6), 752-765.

KEY TERMS AND DEFINITIONS

Business Model: It is a conceptual structure designed as a system to support the methods of business viability, the means to fulfill its purpose and goals, financing, resources, operations, customer base, generation of revenues and profits.

Collaborative Economy: It is a set of initiatives based on horizontal networks with the participation of a community and formed by individuals who have something to share and others who need them with the purpose of giving, swapping, borrowing, trading, renting, and sharing products and services based on distributed trust and decentralized power.

Private Market: The structure of transactions that are negotiated directly between two parties and can take any form the parties agreement.

Strategy.: It is the orientation and delimitation given to resources and capabilities of a firm in relation to complex and uncertain competitive market environments in order to accomplish the vision and mission of the firm.

Technology-Based Company: Also known as tech company, it is a firm focusing on the development and manufacturing that uses leading edge scientific and technological knowledge systematically and continuously to produce new goods or services with high added value.

Transport Incorporated: The act or process of moving people or things through different means of transport from one place to another subject to a patronage or ridership refers to the number of people using a transit unit.

Uber: It is an acronym to mean ultimate, best for above in German. It also has the basic meaning of over, beyond, extremely good. As the trade mark and global brand of a private transportation incorporated company, Uber has been changing inner-city transport structure.

Section 4

Digitalization: Metamorphosis of Educating and Mentoring

Chapter 9
Impact of Digitization on Learning and Opportunities in the Workplace

Vannie Naidoo
University of KwaZulu-Natal, South Africa

ABSTRACT

Digital technologies have permeated all areas of society, be it education, work, business, government or medicine. This has had a major impact on the world of work and its environment as technology continues to permeate all areas of society. There has been a rapid speed of new technological developments in the area of education. These new developments have created learning opportunities for working people. Digital technologies can now assist workers in their training and education needs thus making the worker of today more viable in the labor market. In the workplace, E-learning has made considerable inroads in the lives of workers development. E-learning is a combination of two important constructs namely learning and technology. Learning is a cognitive process where a student filters knowledge that he retains. Technology if used properly in this era can be used as a teaching tool to enable learning.

INTRODUCTION

The world of work has changed. Big companies have to respond to changes in technology in order to remain sustainable and competitive. Whether you are a large company or business operating in any industry the digital component has affected your everyday human resource operations in some way or another. Companies had to

DOI: 10.4018/978-1-5225-3191-3.ch009

take the necessary steps to incorporate the rapidly advancing digital transformation into their strategic framework and business models within all levels of the business. Although an additional cost, stakeholders had to come on board so that the business can survive and thrive in this new advanced era of technology. This chapter will focus on various pertinent issues that surround the impact of digitalization on learning and opportunities for the workers within the organisation.

The following key areas will be discussed:

- How are online courses tailored to suit the needs of working people.
- Perceptions on E-learning in the workplace.
- An overview of On-line Training for staff.
- The advantages of new technology on learning for the working people.
- The disadvantaged of new technology on learning for the working people.
- Strategies to align workers E-learning to the Human resource strategies of the organization.

In the discussion, it follows a historical overview of On-Line learning, E-learning and Life Long Learning will be highlighted and discussed.

Historical Overview of Online Learning, E-Learning and Life Long Learning

Online learning came into being in the early 1980's. If we have to trace the footsteps of E-learning its inceptions can be traced back to conventional distance education or online learning. Moore, Dickson-Deane and Gaylen (2011) argue that the origin of E-learning is not certain. Harasim (2006) argues that the genesis of E-learning as based on human collaboration in knowledge work and innovation can be traced to the development of network communication in the late 1960s, and the invention of e-mail and computer conferencing over packet-switched networks in 1971.

The definition of E-learning is an important starting point in the discussion in this chapter. Hedge and Hayward (2004), defined E-learning as an innovative approach for delivering electronically mediated, well-designed, learner-centred and interactive learning environments to anyone, anyplace, anytime by utilizing the internet and digital technologies in concern with instructional design principles. It is all about learning with the use of computers. Wilson (2001) adds that E-learning is construed in a variety of contexts, such as distance learning, online learning and networked learning. Tavangarian, Leypold, Nölting, Röser, & Voigt (2004) argue that there is no unified or explicit definition of 'E-learning', but its descriptions often emphasizes change in the mode of delivery, such as "a technology-based learning in which learning materials are delivered electronically to remote learners."

Insightful arguments put forward by Khan (1996) is that information and communication technologies offers students considerable benefits including increase access to learning opportunities, convenience of time, and place, making available a greater variety of learning resources, improve opportunities for individualized learning and emergence of more powerful cognitive tools. The most advantageous aspect of E-learning is that it transcends spaces. It offers people at work for example to further their studies without attending class. They can access their E-learning sites from the park during their lunch breaks and do their learning tasks. According to Zhang, Zhao, Zhou, & Nunamaker Jr. (2004) E-learning uses numerous types of electronic media, educational technology, and information and communication technology (ICT) to deliver education. E-Learning provides unprecedented opportunities for people to learn in a more personalized, flexible, and portable manner without the restriction of time and space.

Another concept worth mentioning in this chapter is Lifelong learning. According to Rubenson (2006) and Nicoll & Fejes (2011) the idea of lifelong learning heralds a shift from learning and education activities exclusively bound to the early years of life, towards a continuous learning throughout the lifespan. At work, formal and informal learning is woven into daily practices, thereby separating workplace learning from learning in schools.

In the discussion that follows various perceptions on E-learning in the workplace will be put forward and debated.

Perceptions on E-Learning in the Workplace

The most important resource of any organisation is its human resources. The staff complement of an organisation if trained and developed and motivated accordingly can become the organisations most formidable competitive advantage. Due to time constraints at work, work pressure or family commitments, opportunities to study and upgrade their qualifications were often not possible for an organisations employees. Certain people at work sometimes were previously not allowed to study. An example is in certain traditional homes women were not allowed to attend university. However, through E-learning, a woman who was previously disadvantaged can now study and receive a degree or certificate. The E-learning platform has changed the dimensions of teaching and learning. In a traditional system, the learners had to be present in the classroom. In E-learning, the learner can access the learning site at home, in the park or even at work. There is no more space constriction.

Education empowers people by giving them critical skills and means to create opportunities for sustainable and viable personal and economic growth. One of the most significant changes in educational practice is the shift from traditional teacher-centred pedagogy to learner-centred pedagogy (Sandholtz, 1997). Creutz and

Wiklund (2014) indicate that there is an increased use of E-learning in training of employees which is claimed to provide new learning possibilities and implications for employees and the organization. Rubenson (2009) and Cropley (1979) have pointed out that workplace learning has become an important constituent in the process of learning throughout the life span. Hence, learning is considered to enhance economic progression, individual and personal development, and increase social balance in a globalized knowledge economy. (Hughes, 2010) add that there has been some debate on how emerging technologies may enable pedagogically sound training initiatives, in which there is a greater focus on the learner's experience, allowing for a feeling of learner ownership, alongside reciprocity, dialogic reflection and a processual focus.

Some other interesting insights put forward by theorists are that: "Technology provides one answer to the design of effective workplace learning environments. E-Learning: make learning anytime, anywhere a reality which particularly appeals to the corporate world because it allows workers to learn constantly without time and/or geographic boundaries" (Wang, 2002). Hrastinski (2008) adds that in his has studied asynchronous learning methods are used in E-learning educational setting. It stands to reason that E-learning in the workplace also takes place through asynchronous learning methods under the label of flexibility for employees and production.

E-Learning covers a wide set of applications and processes, such as Web-based learning, computer-based learning, virtual classrooms, and digital collaboration. It includes the delivery of content via Internet, intranet/extranet (LAN/WAN), audio-and videotape, satellite broadcast, interactive TV, and CD-ROM. E-learning can be delivered through synchronous and asynchronous means. Synchronous E-learning models tend to follow the traditional-classroom model. Online lectures are the "driver" of the course (Midkiff and DaSilva, 2000 in Kakkar, 2008). The asynchronous E-learning model requires workers who are students in the program to learn on their own.

According to American Society for Training and Development (ASTD, 2010), almost one third of training for employees in the workplace has been implemented through technology. Bersin (2005) added that in his survey obtained from 526 companies in the USA and Canada, E-learning was on the rise and continued to grow in 2005 by 25%. Welsh et al. (2003) adds that "the future of E-learning is in the integration of information management (including training content material, the learner's current knowledge, and the learner's training activities, often now distinctly referred to as content management, knowledge management, and learner management systems), performance support, peer collaboration, and training systems. For example, in the future, a sales employee should be able to use a single intranet portal to collect information about potential customers, find a quick answer to a customer query, interact with other sales staff throughout the country, and take a class about sales techniques."

In recent years, with the growth of corporate E-learning, proper learning of the personnel in charge of E-learning and efficient linkage with systematic knowledge management have emerged as significant components for intellectual management and improvement of business performance (Ertl, 2010). Investments in employee learning have been shown to benefit organizations' innovative performance (Sung & Choi, 2014). Kyndt, Govaerts, et al. (2012) argues that the core factor in providing motivation for employees to engage in learning activities is as follows -

- The degree of autonomy.
- The content and complexity of the job.
- The learning potential.
- The task variety.
- The mobility opportunities the job offers.

There is no one E-learning system that fits the requirements of employees at all organisations. Organisations are unique and their employees are complex with needs, attitudes and motivation that is unique to another organisation. What motivates and drives one employee may not necessarily drive another although they work within the same organisation. Caudill (2015) argues that this systematic approach to workplace E-learning, from the first steps of matching the learning environment to learner motivations and establishing a learning culture to assessing the impact of the E-learning efforts on organizational performance makes the application of E-learning in the workplace unique. Naidoo (2017) adds that the benefit of having an E-learning system is that learners from all age groups can work and study at their own time. This flexibility is a benefit to many who are illiterate or semi-literate who can work and study.

The E-learning system and what to teach its employees also is an important consideration to take into account as the organisation needs to have control over its strategic knowledge assets. This is why a not a lot can be gaged by studies on employees E-learning training as a company may view its employees and E-learning training system as a source of its competitive advantage that must not be shared by its rivals in the same industry. Alvesson and Kärreman (2001) add that a pertinent question, in fact, is whether formal learning and training designers in organizations can actually afford to relinquish control to learners, given the discourse of knowledge-based competitive advantage of corporations, alongside the institutionalization of benchmarking within specific industries.

After an in-depth discussion on perceptions on E-learning in the workplace the next theme that will be highlighted is an overview of On-line training for staff.

An Overview of Online Training for Staff

Training staff is one of the most important strategic goals of an organisation. By training staff to be more efficient in their job performance would ensure higher productivity. Better trained and developed staff can also become a major internal strength or competitive advantage of an organisation. Staff train and develop themselves because they believe strongly that this makes them employable and a lucrative asset to the organisation. According to Pride, Huges and Kapoor (2017) the following are training and development methods available to employees within the workplace -

- **On-the Job Methods:** The trainee learns by doing the job tasks under the supervision of an experienced employee. For example, when the University of KwaZulu-Natal bought a new software package to detect plagiarism. Academics staff had to be trained by a trained professional from the university's IT division on how to use this software to detect plagiarism in student's work at masters and Ph. D. work.
- **Simulations:** The work situation is simulated in a separate area so that learning takes place away from day –to-day work pressures. Project managers in construction companies and engineers use this when constructing a bridge for example; they run simulations with their teams so that they can be precise in their building construction. If they are not precise and off a few centimeters it can compromise the project. Simulations is a form of learning as well as a good way for project managers (Engineers) to be able to train the trainee project managers on how to construct bridges for example in major projects for the company.
- **Classroom Teaching and Lectures:** Instructors present the employee with concepts and illustrations through a variety of techniques. At the previously known University of the North QwaQwa campus trained professionals from the National Research Foundation (NRF) were called in by the university to come to the campus and teach junior academics on how to draw up their Masters/ PHD research proposals and design their questionnaires and other research instruments for their chosen studies. The learning opportunities took place at the place of work. Cost were saved by the university as a trained phycologist and senior researcher conducted these workshops at the campus for 12 staff for example and staff did not have to travel to the NRF's office in Pretoria to receive training.

- **Conferences and Seminars:** Experts and learners meet at these conference or seminar where they exchange and share ideas. Many medical conferences occur each year for example. This is where expert's in the field of medicine in specific areas who are doing ground breaking research are brought together with other medical professionals from different hospitals and medical organisations. The training medical professionals can listen to these seminars and conferences and learn and employ it within their organisation.
- **Role-Playing:** Participants act out their other's roles in the workplace to better understand them (this is a primary management development tool). Within corporate organisations roleplaying is an excellent training tool to assist staff in understanding group dynamics and being in reverse roles to their managers.
- **E-Learning:** Employees are trained by watching videos of lectures or how-to-guides, "game simulations" that simulate a work situation using a computer, or taking on-line quizzes to demonstrate the workers proficiency in a topic. An interesting example is, at some universities the medical faculty uploads live feed of medical procedures so that medical interns can access this on the Moodle sites for E-learning together with the podcast of the days lecture. Learning therefore occurs outside a traditional classroom.

According to Nisar (2002), E-learning allows an efficient response to the following specific objectives: (1) to identify and record training needs of workers, so as to facilitate the development of ad-hoc training processes; (2) to provide up-to-date training and support to workers who need it, allowing a personalized interaction between user and tutor; (3) to evaluate through digital technologies the established training objectives, via a process of continuous assessment in order to facilitate the assimilation of knowledge acquired by the worker; and, finally, (4) to establish a registry and record of training activity and of each participant's assessment. Park and Wentling (2007) claim that: "When people are faced with a computer that is the primary tool in E-learning, they might show various kinds of attitudes and a wide range of self-efficacy toward computers. It has been demonstrated that these attitudes and self-efficacy influence the results of any activities completed via computer and it has been shown that self-efficacy has either a direct or indirect relationship with training outcomes in general".

Efficient and effective training methods are always the key for companies ensuring that their staff and partners have the latest information and instructions. Harrying to meet this need, universities and commercial entities around the world offer thousands of online courses, including certification and higher education programs. For example, in 2001 the Massachusetts Institute of Technology promised to freely publish all their training materials for non-commercial use. A year later, nearly 50,000 students, about 70% more as in the previous academic year enrolled the first and second stages of higher education at on-line University of Phoenix-e (Shea, 2002).

Overton (2004) conducted studies involving 16 European organisations with 2,000 learners and the results indicated that these organisations were recognised for their sustained success in embedding E-learning within those organisations. Sloman and Rolph (2003) highlighted that factors that impacted on workplace E-learning included strategic intent, effective introduction, blending of learning modes, learner support and effective measurement/monitoring. Peter Guiney (2015) commented that Energy Australia is reported to have saved AU$100,000 per annum in staff training costs because e-learning reduced the frequency of workers having to attend their training centre, and decreased significantly the time taken for course delivery.

According to David, Salleh and Iahed (2012) if you are using E-learning training in the workplace the following is expected -

- Interactive sessions at the computer with other participants and E-learning facilitators who are online, or communicating via phone.
- Completing activities at the computer that are available from a structured learning site, designed specifically for the course.
- Downloading (and possibly printing) course material to read.
- Researching information from other websites.
- Listening to a podcast.
- Viewing a YouTube video or similar.
- Posting comments to an online chat forum, such as a wikispace.
- Completing an access task online.
- Attending a face-to-face session facilitated by a trainer to further investigate course content and concepts.
- Discussing content or seeking advice from colleagues at work in relation to the course material, David, Salleh and Iahed (2012).

According to learning activities in the modern firm can encompass a number of different areas. Technical learning involves developing new skills for new tools, whether those skills are applied to a new software application, new manufacturing equipment, or some other technical aspect of the business. This type of learning is often the first type considered because of the rapid advance of technology in all aspects of life but it is not the only area of change and learning for today's workers.

In the discussion that follows the advantages of new technology on learning for the working people will be put forward.

The Advantages of New Technology on Learning for the Working People

There are various advantages associated with using ICT's in the workplace as an opportunity to develop staff. E-Learning can also be used by the organisations staff to further credential them with the necessary qualification that can allow them to move into higher positions within the organisation. The major benefits of E-learning will be highlighted below -

- E-Learning is flexible to access. People who work shifts like police men and nurses for example have different styles of working and learning. E-learning is flexible to access and can be adapted to suit the lifestyle of employees. This way the employee chooses when to learn and he/she can always re-visit the E-learning site at their convenience.
- E-Learning is not restricted to boundaries and space. The employee can visit the E-learning portal both on –site in the company or off-site in at home or in the park.
- E-Learning is a very good learning tool to those employees who are quiet by nature or shy. In a traditional classroom, employees such as these would feel intimidated but on-line they are anonymous and can share their views with their instructor and others on the learning site chatrooms.
- If the E-leaning site is designed to be interesting and interactive then employees would be motivated to learn. By adding videos, voice prompts, an E-tutor makes the E-learning process more interesting and keeps the employee who has had a busy day interested in the subject matter.
- If the employees are trained using E-learning on site within the organisation, this would save the company on time. For example, the training is on-site for 1 hour and employees then can return to their duties after and overall productivity within the workplace will not be lost.
- E-Learning is very attractive to the young people in today's world who are very techno-savvy. If these employees were given the opportunity to train and develop their skills or expertise they would take up the opportunity as they have the know-how to navigate the E-learning websites quite easily than the older generation.
- If the organisation has well skilled management and ICT staff in place then training using E-learning within the organisation would be less costly and less time consuming especially when there are problems encounter during the training.
- If companies provide E-leaning opportunities to their staff they would feel appreciated and would be more motivated to stay at the company.

In the discussion above the major advantages of new technology on learning for the working people was highlighted. In the discussion below the disadvantages of new technology on learning for the organisation and its staff would be put forward.

The Disadvantages of New Technology on Learning for the Organisation and Its Staff

The following are some barriers to E-learning in the workplace -

- The cost of implementing E-learning systems is expensive and organisations often grapple with convincing their shareholders and directors to make such investment.
- Lack of infrastructures is another drawback. Example in Africa where there are many power outages it is often difficult for learners to connect to the E-learning sites.
- Management is too focused on the bottom line, they refuse to develop and upgrade their staff's skills because they want to please the shareholders and retain their high bonuses and continued seat in an executive position within the organisation.
- Staffs are not willing and show great reluctance towards E-learning or going on ICT training.
- Staff reluctance to change and improve themselves, they are from another generation and refuse to adapt to the digital age.
- Management lack the training to facilitate E-learning training and development within the organisation.
- If employees' lacks family and work balance it is often difficult for them to embark on E-learning opportunities to gain certification or training.
- If staffs are demotivated with the present management structure they will not be persuaded to commit to E-learning to further advance their person growth or improve their skills and expertise to be better performers within the workplace.
- If previously staff was involved in E-learning to train and advance themselves and they had a bad experience or they perceived the quality of E-learning as poor they will not be attracted to continue. They will even tell other employees of their bad experience and word of mouth carries great persuasion not to use the E-learning platforms to train or develop them in the workplace.
- E-Learning in the workplace is a great strategic tool however if there is insufficient computers in the workplace or staff do not have access to computers they will not be able to use this technology.

In the discussion below strategies to align workers E-learning to the Human resources strategies of the organisation would be highlighted.

Strategies to Align Workers E-Learning to the Human Resources Strategies of the Organisation

The main strategic objective of any organisation is to remain competitive within the industry in which it operates. In order to do so it has to have the right staff complement that can implement its various organisational, departmental and operational strategies. The main purpose of the human resource executive is that it works to develop and train its workforce so that they are productive in their job performance. A useful strategic tool is E-learning that can be used by the executive management in the department of human resource to successful train and develop their employees. At the end of the day a good E-learning system in the workplace depends on the organisation and its leadership and employees who are motivated to succeed. McPherson and Nunes (2006) categorized the organizational critical success factors of E-learning largely into leadership and cultural issues, design issues, technological issues and delivery issues and reported that the leadership and cultural issues should be crucial factors determining the success or failure of change and innovation.

In order to align workers E-learning to the Human resources strategies of the organisation entails the following -

- There must be clear strategies in place for training and development of staff using E-learning. An example could be that the organisation develops policy documents on E-training and E-learning. This gives direction to staff on the digitized learning opportunities available to them.
- It is often important to have an effective top management team in the human resource department that champions E-learning in the workplace for staff. The executive management team must be passionate about E-learning initiatives to drive them throughout the organisation. If the executive management in human resources is old school and does not see the gains of E-learning in the workplace or they are too frugal and refuse to spend on money to train their staff then they would not support such initiatives within the organisation.
- There must be effective communication within the different levels of the organisation to make staff aware of E-learning opportunities within the workplace. Through proper e-mailing staff can be notified on E-learning opportunities. The Company website can also be regularly updated to help communicate to staff the training opportunities available to them in the year.

- The manner in which staffs are selected for E-learning training within the organisation should be done in a fair and equitable manner so that staff does not feel left out or victimised by the process of selection. Staff with the necessary skill and potential need to be selected on training E-learning for example. You cannot have the tea lady go for computer training because she is your friend. It is important to reward employees who are excelling to go on advanced E-training programs so that they can add value to their jobs and make them even more competent in their performance of their duties.
- The executive management and his/her team who drive E-learning training opportunities within the workplace must be well skilled in ICT's. With the advancements in technology occurring at such a rapid place worldwide. To remain competitive in industry it is advisable for the company to employ executive management who are competent and well skilled in ICT's so that they can properly champing E-learning opportunities in the workplace.
- Proper ICT infrastructure and ICT support staff must be present in order for E-learning training sessions to go smoothly. Many companies invest large sums of money on E-training for example within the workplace without having proper infrastructure in place or training ICT staff support. Many E-learning attempts first failed at schools and universities in Africa for example, because staff had no training on how to use the complex computer software. Another reason for failure was students or staff had no access to internet because of the country's poor infrastructure since some countries in Africa experience large power outages and had no proper broad band internet access as their Western counterparts.
- E-Learning in the workplace should be aligned to the Human resource strategies so that they can support it each and grow the staff complement through training and development in order for the organisation to maintain competitiveness in the industry. Well trained staffs are worth a fortune to retain as he/she knows their job in an efficient and effective manner. If people are looked after in an organisation, they will be loyal and committed and they will always serve the organisation well and this will be positively received by the organisations customers. Apple is a very commendable company as it spends a lot of money in training of their staff who they value as their most prized asset.

FUTURE RESEARCH DIRECTIONS

E-Learning opportunities in the workplace in developing countries should be the focus of future research. Since there is a gap in how developing countries are involved in E-learning and E-training and to what extend and how do their staff perceive E-learning opportunities are very important questions that need answers.

Another area for future research is to compare company policies on E-learning opportunities and see what different countries in the world are doing to actively train and develop and retain their most prized asset, their staff.

CONCLUSION

In the 21st century technology has become the cornerstone of society and corporate entities. Businesses can no longer survive without emplacing new technology in the work environment. The digital divide that once divided the many counties in the world is now brining everyone in the world closer. In this global village employees have to be trained and developed so that they can be efficient in their jobs as the new technology are being used extensively by businesses and corporates throughout the world. Human resources are the sector within a company that has the most change as digitalisation affects their staff in some way or another. There are opportunities available via E-learning and distance education that can assist the employees to upgrade their skills and qualifications. Digitisation and advancement in technology can now open new doors to staff that were unable to attend university. Training can even be done on the job using new technologies, where staff can be trained on-line without leaving the office. E-Learning is just a teaching tool that can assist employees to be trained and developed within the workplace. There is no one solution that fits all in E-learning within the workplace, as all employees within an organisation have their own unique personality, lifestyle and attitude. E-Learning will continuing to re-define the workplace and it is up to the organisation and the employee to see the opportunity this system of learning proves and take advantage of it.

REFERENCES

Alvesson, M., & Kärreman, D. (2001). Odd couple: Making sense of the curious concept of knowledge management. *Journal of Management Studies*, *38*(7), 995–1018. doi:10.1111/1467-6486.00269

American Society for Training and Development (ASTD). (2010). *State of industry report: Executive summary. Alexandria*. Alexandria: VAASTD Press.

Caudill, J. G. (2015). Employee Motivations for Workplace Learning and the Role of E-learning in the Workplace. *Internet Learning*, *4*(2), 37–48. doi:10.18278/il.4.2.3

Sloman, M., & Rolph, J. (2003). Change Agenda. *CIPD*. Retrieved 5 September 2009, from http://www.cipd.co.uk/NR/rdonlyres/2716D932-3FEF-4559-9667-2AD7006EA13F/0/elrn_curve.pdf

Cropley, A. J. (1979). *Lifelong education: A Stocktaking*. Hamburg: UNESCO Institute for Education.

Ertl, B. (2010). *E-Collaborative knowledge construction: Learning from computer supported and virtual environments*. Retrieved from http://www.library.uiuc.edu/proxy/go.php?url=http://services.igiglobal.com/resolvedoi/resolve.aspx?doi=10.4018/978-1-61520-729-9

Guiney, P. (2015). E-learning in the Workplace- An annotated bibliography. New Zealand: Ministry of Education. Retrieved from www.educationcounts.govt.nz

Harasim, L. (2006). A history of e-learning: shift happened. In J. Weiss, J. Nolan, J. Hunsinger, & P. Trifonas (Eds.), *The international handbook of virtual learning environments*. Dordrecht: Springer. doi:10.1007/978-1-4020-3803-7_2

Hedge, N., & Hayward, L. (2004). Redefining roles, University E-learning contributing to Life-long learning in a networked world. *E-learning*, *1*(1), 128–145.

Hrastinski, S. (2008). Asynchronous & synchronous e-learning: A study of asynchronous and synchronous methods. *EDUCAUSE Quarterly*, *31*(4), 51–55.

Hughes, J. (2010). But it's not just developing like a learner, it's developing as a person – Reflections on e-portfolio-based learning. In R. Sharpe, H. Beetham, & S. De Freitas (Eds.), *Rethinking Learning for a Digital Age: How Learners Are Shaping Their Own Experiences*. New York: Routledge.

Khan, A. (1996). Utilization of communication technologies for distance. In K. Murali Manohar (Ed.), Distance education theory and practice: Media and Communication Technology.

Kyndt, E., Raes, E., Dochy, F., & Janssens, E. (2012). Approaches to learning at work: Investigating work motivation, perceived workload, and choice independence. *Journal of Career Development*, *40*(4), 271–291. doi:10.1177/0894845312450776

McPherson, M., & Nunes, M. (2006). Organisational issues for e-learning critical success factors as identified by HE practitioners. *International Journal of Educational Management*, *20*(7), 542–558. doi:10.1108/09513540610704645

Moore, J. L., Dickson-deane, C., & Gaylen, K. (2015). e-Learning, online–learning, and distance learning environments: Are they the same? *The Internet and Higher Education*, *14*(1), 129–135.

Naidoo, V. (2017). E-learning and Management Education at African universities. In N. Baporikar (Ed.), *Management education for global leaders*. USA: IGI Global. doi:10.4018/978-1-5225-1013-0.ch009

Nicoll, K., & Fejes, A. (2011). Lifelong learning: A Pacific Know How. *Studies in Philosophy and Education*, *30*(4), 403–417. doi:10.1007/s11217-011-9235-x

Nisar, T. M. (2002). Organisational Determinants of E-learning. *Industrial and Commercial Training*, *34*(7), 256–262. doi:10.1108/00197850210447246

NorénCreutz, I., & Wiklund, M. (2014). Learning paradigms in workplace e-learning research. *Knowledge Management & E-Learning*, *6*(3), 299–315.

Overton, L. (2004). *Linking learning to business: research report 2004. Reading*. UK: Bizmedia.

Pride, W. M., Huges, R. J., & Kapoor, J. R. (2017). *Foundations of Business*. Singapore: Cengage Learning.

Rubenson, K. (2006). The Nordic model of lifelong learning. *Compare: A Journal of Comparative Education*, *36*(3), 327–341. doi:10.1080/03057920600872472

Rubenson, K. (2009). Lifelong learning: Between humanism and global capitalism. In Jarvis (Ed.) The Routledge International Handbook of Lifelong learning. Routledge: London & New York.

Shea, R. H. (2002). E-Learning Today; As an industry shakes out, the survivors offer no frills education for grown-ups. *U.S. News & World Report*, *133*(1), 54–56.

Sloman, M., & Rolph, J. (2003). *E-learning the learning curve: the change agenda. Institute of Personnel and Development*. London: IPD.

Sung, S., & Choi, J. (2014). Do organizations spend wisely on employees? Effects of training and development investments on learning and innovation in organizations. *Journal of Organizational Behavior, 35*(1), 393–412. doi:10.1002/job.1897 PMID:25598576

Tavangarian, D., Leypold, M. E., Nölting, K., Röser, M., & Voigt, D. (2004). Is e-learning the Solution for Individual Learning. *Electronic Journal of E-learning, 2*(2), 273–280.

Welsh, E. T., Wanberg, C. R., Brown, K. G., & Simmering, M. J. (2003). E-learning: Emerging uses, empirical results and future directions. *International Journal of Training and Development, 7*(4), 245–258. doi:10.1046/j.1360-3736.2003.00184.x

Wilson, J. (2000, February). Lessons of a Virtual Timetable: Education. *The Economist.*

Zhang, D., Zhao, J. L., Zhou, L., & Nunamaker, J. F. Jr. (2004). Can e-learning replace classroom learning? *Communications of the ACM, 45*(5), 75–79. doi:10.1145/986213.986216

KEY TERMS AND DEFINITIONS

Competitive Advantage: This is where an organisation has an edge over their rivals in the same industry.

Credentialing: This is when staff gains an accredited qualification that can assist them in their job. Example when the organisation's trainee accountant gets his B. Com (Accounting) degree.

Digitization: It refers to data being converted to digital form to be used in computers, smartphones, laptops, videos, text messages or photographs.

E-Learning: This refers to computer based learning that staff in the workplace can take advantage of.

E-Training: This refers to computer based training taking place in the workplace.

Skills: A skill can refer to how an employee carries out his tasks in the organisation. Skills can refer to intellectual skills, computer skills for example.

Techno-Savvy: This means you good at using the new technologies. For example, if a young staff member is good at using computers, smartphones and other mobile devices.

Workers: This refers to the staffs who are employed by the organisation to do a specific job.

Chapter 10
Social Media as Efficacious Tool for Teaching

Swati Panchal
Mayo International School, India

Lakshay Panchal
Netaji Subhas Institute of Technology, India

ABSTRACT

Social Networking is a good source to stay connected with others so it can be used as a tool in teaching. Those students, who are not able to learn the concept in the regular classroom, go to tuition. They work whole night to complete their homework and assignments meanwhile they cannot resist themselves from using social networking sites. They have a view that they can multitask, but actually, most of the concentration is in social networking and messengers and less in studies. So ultimately their studies suffer. Because of this, they feel unrest in next day classroom also. To change the habit of using the mobile phone and social networking is very difficult. Now they are addicted to using these. So, the only solution is to change the way they use these things. The only requirement is training the student about the use of these systems, for which they are made of. So, the basic aim of this research is how social networking can be used as an Efficacious tool in teaching and utilizing the time wastage on social networking sites and messenger by students in studies.

INTRODUCTION

In the modern smart world, everyone is enjoying the togetherness, everyone is enjoying the oneness despite the distances, despite the cultural and territorial boundaries, everyone is cheering for one another, everyone wishes their near and dear ones, everyone is talking to another person, but yet a word has to be spoken. So much of things communicated in and from every corner of the world but not a single word

DOI: 10.4018/978-1-5225-3191-3.ch010

is voiced by all. Everything is said and expressed by without any expression and any oral communication. Physically, all are present and communicating but with complete silence. Everyone's heads are down looking at their phones, tabs, etc. It's an ugly truth of today's world where everyone is virtually connected to far away from people, but they are going far from the people near to them.

Thousands of years ago, when civilization was developing, one important thing that emerged without which today's world cannot be imagined, was the 'Art of Communication.' At the start, basic pictorial representation was used, and as the time passed by, various communication mediums were developed. Languages were developed, and then documentation of the thoughts was also done by writing them on rocks and leaves. After that, the paper was prepared, after that especially for long distance communication mobile phone was developed. Now a day's mobile phone is not only used for communication, but it serves many other purposes.

In this 21st century and late 20th century, the mode of communication took a major leap forward and adopted the new innovative ways to communicate. Now the present scenario can be perfectly described as "…the human's voice noises are now mostly replaced by the digitally propagating frequencies." Now the point to notice or to bother about is that this change is not entirely physical rather it's also slowly transforming and bending the human's mental and psychological behavior too. The above statement can easily be justified by the fact that now mostly people have started preferring online meets and interactions over face-to-face meet; texting is now being preferred rather than talking to one another.

Therefore, one can say that people are becoming more social and more interactive with others and the style of communication has changed so much that we are now not meeting face-to-face as we used to do earlier. Also, according to the studies, the interactions we do virtually or online are the ones in which we don't feel personally connected, or any attachment level is lacking as compared to the case when we interact face-to-face. Thereof, we can say that by using these social networking sites, we all are making many bonds and ties with too many people but those ties and bonds tend to be much weaker in the true sense.

In addition to it, the feature to choose individual's society and circle on by own comes as a boon as well as a bane. Whereas it provides a relatively suitable environment for the introvert kind of people while on the other hand, it confines one's social circle having the same views and thinking and eventually hampers the openness and exposure to diversity. All these things are making the world a virtual global village with some societies and groups within itself where the overloaded raw information travels with the speed of light and spreads to almost the whole audience. This raw, unverified information creates a very wide and complex behavioral impact on the audience. Sometimes these things spread so widely and virally that they set the trend in the mass, mostly of the slings (that are now being accepted and welcomed by society too) are the result of these social media trends. This ability to communicate with more people across greater distances and with increased speed makes the social media as one of the most useful tool (or one can say weapon) for many societies and individuals. The matter of fact is that the message can be shared and spread farther and faster than it has been at any other time in the human history.

Social media will likely continue to become increasingly integrated into the normal human experience like most of the communication technologies that preceded it. They will continue to increase the volume of the human communication process, and we will continue to learn how to use them for good and for bad too. We must not forget that social media is only a decade old, and that's very young in the history of communication technology wherein it has been influential, but it hasn't settled into a routine yet. And it's quite possible that we may rely on the weak connections we're making on social media, but despite these, there are some advantages of social media just like two sides of a coin i.e. if it can make us weak, it can make us strong too.

Communication through direct or by using social networking limits economic, social and political boundaries. Social media gives a platform on which people can share news, views, and opinions regarding the happenings around them. Social media is not only used for communication and sharing, but it is also being used by the advertisers for promotions, recruiting, as social learning tools, etc. It provides a simple way to let students, parents, and educators collaborate online. An example of social media is blogs, websites, and applications.

In collaborative technologies where blogs and social networking websites lead to instant online communication in which people communicate more rapidly and conveniently with each other. Nowadays, social networking is rapidly growing all over the world and India is one of the biggest growing social networking countries. Social networking is what started in the past as a hobby for few computer literate people; now it has become a social norm and social status for all. Teenagers and youngsters have started using these sites very roughly without any rational thinking wherein they connect with peers and other people, share information and showcase their social lives explicitly. Presently, their main motive is to get increased number of views, likes, and comment but if this habit changes, this will become a positive point for learning, and it can give self-motivation to students as well. Students can't learn by only writing on a piece of paper, and so it is also important to involve them practically such as practical work, blogging or any other way in which more sense organs can be used. Blogs are the websites where they can write not only for the teachers or educators but for themselves to enhance their skills.

WhatsApp and other mobile messengers, Facebook and other social networking websites, Gmail and other email providers, YouTube, and other video websites are also there which are used for communication too. These all are used widely but seen in a negative scene these days, the reason being that these tools are not bad, but the way they are used is bad like they are used for chatting on irrelevant topics, posting the day to day activities, etc. But there is another way to use these tools like by posting articles, posting views on a social cause, puzzles, pooling, etc. These are the benefits provided by these tools, but the point is that the students don't use them in a needful manner. If these tools are included in the curriculum and training is given to them, this will become their habit and will help them to learn more and grasp more. If the use of these tools gets changed, they will surely become the smartest tools in the learning process.

Some of the popular social media websites are -

- **Facebook (FB)** is an online social media and social networking service. Facebook (FB) may be accessed by a large range of desktops, laptops, tablet computers, and smartphones over the Internet and mobile networks. Users can add other users as "friends," exchange messages, post status updates and digital photos, share digital videos and links, use various software applications ("apps"), and receive notifications when others update their profiles or make posts. In groups, editors can pin posts to top. Facebook (FB) was the most popular social networking site in the world, based on the number of active user accounts. Facebook (FB) classifies users from the ages of 13 to 18 as minors and therefore sets their profiles to share content with friends only.
- **WeChat** is a free, cross-platform and instant messaging application developed by Tencent. As of May 2016, WeChat has over a billion created accounts, 700 million active users; with more than 70 million outside of China (as of December 2015). In 2016, WeChat has currently 864 million active users.
- **YouTube** is a video-sharing website. Most of the content on YouTube has been uploaded by individuals. Unregistered users can only watch videos on the site, while registered users are permitted to upload an unlimited number of videos and add comments to videos.
- **Twitter** where registered users can post tweets, but those who are unregistered can only read them. Twitter proved to be the largest source of breaking news.
- **Snapchat** is an image messaging and multimedia mobile application created by former students at Stanford University and developed by Snapchat Inc.
- **Skype** is an application that provides video chat and voice call services. Skype allows the creation of video conference calls.
- **WhatsApp** Messenger is a freeware, cross-platform and end-to-end encrypted instant messaging application for smartphones.
- **Tango:** The app is free and is popular for offering video calls over 3G, 4G, and Wi-Fi networks.
- **Wikipedia** is the free online encyclopedia. There are 5,349,007 articles on the site.

- **BlackBerry Messenger**, also known as BBM.
- **LINE**
- **Viber**

These are the best ways to reach the people. Since human nature is to get attracted towards a lively environment, and so they find books to be dull and boring as there is no animation in them, which cannot provide the liveliness so desired. Social networking sites, Messengers, and YouTube attract students as these tools provide the live environment and give the student an environment in which they can interact with others. Nowadays, these tools are used most frequently but for communication and entertainment purposes only. If these tools are used in teaching and learning, these tools can help a child to understand a concept in a better way. A child can play – pause-replay it as many times as he/she wants to and can even ask questions from the teachers when he gets stuck up. This process helps both the teachers and the students equally. The teacher can suggest a link to a video of a difficult topic and a student can visit the video as many times as he wants. In this way, the tuition business or coaching institute can be controlled from which both teachers and governments are suffering. The technology that can be used in teaching is Information and Communication Technology i.e. ICT.

Information and Communication Technology (ICT) in Teaching Learning: Information and Communication Technology (ICT) provides a variety of content like animations, video lectures, videos, cartoon movie, etc. which help learners in concentration, better understanding and long retention of information which is not possible otherwise. Websites that are available freely may be used by teachers and students to understand different concepts, improve vocabulary, develop reasoning & thinking, etc. There are websites on which teachers can create their content so that the students can access it at home when they are absent or not able to attend the class.

Use of Information and Communication Technology (ICT) in Developing Virtual Laboratory: The students understand better if they do some practical related to the concept as it makes learning easy and interesting. A Laboratory helps in developing scientific temper. But the fact is that practical are not performed by each student in every school. There are many schools which do not have a laboratory. Sometimes if the laboratory is available, the instruments are not available. The students do not have freedom to do experiments on their own. Then the different websites can be suggested or provided which help them to do practically in a virtual lab. An example of this is www.olabs.co.in.

The Computer is an electronic device that can store, retrieve & process both qualitative & quantitative information fast and accurately. But researchers started using Computers for teaching purpose. It gave birth to Computer Assisted Instruction (CAI), Computer Managed Instruction (CMI), Computer Based Instruction (CBI), etc.

Mobile is also presently being readily used for teaching and learning process. The applications which run on the smartphone are great if they are used properly. Technology only gives us a helping hand it is our matter of choice, how we use it.

Information Technology (IT) is used to develop websites which open up a new source of information. The government, corporate sector, educational institutions, etc. have started uploading the information on their websites on instant basis. It also provides facilities for Chat, e-mail, surfing, etc.

The Information and Communication Technology (ICT) = Information Technology (IT) + other media. It has opened new ways of online learning, e-learning, Virtual University, e-coaching, e-education, e-journal, etc. Third Generation Mobiles are also part of Information and Communication Technology (ICT) and can be used for imparting information in the fast and cost-effective way. It provides e-mail facility also which one can access anywhere and at any time too. The Information and Communication Technology (ICT) brings life to the classrooms and libraries of the teachers and students. It provides the prospect for the learner to utilize maximum senses to dig up the information. It can be used both at school and higher education levels in Teaching, Remedial Teaching, Diagnostic Testing, Virtual Laboratory, Online Tutoring and Development of Reasoning & Thinking Skills, Instructional Material Development, Evaluation and Psychological Testing.

Mostly teachers undertake one-way teaching method in the classroom, but while doing so, they forget that teaching is not only imparting subject knowledge but it is also about the real-life application of that concept. Thus, along with subject content, the teacher should:

- Help to develop understanding in concept and application of the concepts,
- Help to develop reasoning and thinking ability,
- build decision-making ability,
- Help to develop self-concept, and value, and
- Help to develop tolerance and ambiguity, risk taking capacity, scientific temper, etc.

Use of Information and Communication Technology (ICT) in Online Tutoring: It is a well-known fact that all students do not understand all subjects to the same extent and even some students find subjects like Mathematics, Physics, English, Chemistry, Accountancy, etc. difficult. Therefore, students go for tuitions. In Online tutoring, the student stays at his home and logs into his tutor through the use of internet and software.

Now, he can watch the teacher who is in one place, and the teacher can see the student who is some other place. The student can ask questions, and the teacher can reply to it by writing on the soft board or by using power point presentation. Teachers should provide, to the extent possible, free of cost WhatsApp and Messenger support to the students.

Teaching learning process of every student or learner is very important. Given are some basic questions while selecting a teaching learning material –

- What should he or she learn?
- From where should he or she learn?
- Is that method reliable for learning?
- How is content presented in front of the learner?

A teacher can use any of the following methods in teaching to impart knowledge to the students -

1. **Instructor / Teacher Centred Method:** In this method, a teacher plays a role of a master as he is supposed to be a subject matter expert and the learner is supposed to be a recipient only as there is no space for student's thoughts. Since there is no involvement of the learner, he has to sit idle and listen to the teacher which means that in this method of teaching, an element of live environment is missing.
2. **Learner Centred Method:** The learner and the teacher interact with each other in a live environment, and the learner is the prime focus in this method of teaching. The teaching strategies are developed in such a way that what learner already knows is also duly discussed. The teacher works as a mentor, and in the process of discussing numerous things in this method, the teacher himself also learns many new things. The learner can ask or inquire about the topic or problem, and an open discussion is welcomed under this method of training.
3. **Direct Method:** In this method, the teacher decides what is important for the learner, and then he demonstrates the skill for the learner to repeat it. Very less abstraction is involved under this method of training and is frequently used in high school practical sessions.
4. **Drill and Practice Method:** This technique is of the nature of rote memorization and is used by students for a particular skill learning like to teach multiplication table, etc.

5. **Lecture Method:** A Lecture is nothing but presenting the content without exchange of idea between a teacher and the learner. A Large amount of information of the content is delivered to a large group of students in a very small time. Therefore, lecture method of teaching is a one-way learning method.
6. **Question and Answer Method:** Students ask questions from the teacher, and the teacher responds accordingly, and vice versa can also be possible in this method of teaching. A student uses his prior knowledge to answer the question asked by the teacher. Also, a teacher can also know the opinion of the student on the given topic and check their level of understanding. Under this method, a student formulates or reformulates their ideas without diminishing the value of the original idea.
7. **Discussion Method:** A teacher and student in this method are on the same stage, and none of them holds an upper hand in the discussion. This method of teaching involves an exchange of Ideas, and they may agree or disagree over with the idea presented according to their levels of understanding.
8. **Discovery Learning Method:** This method of teaching focuses on the student's personal experience and foundation for conceptual development. A student is asked to find information on a given topic and then share their common experience in class.

BACKGROUND

Social networking is not new to the world it was introduced in 1997. Sixdegrees. com was the first social networking website which was launched in 1997. The user can list profiles and can contact them, but got closed in only three years (Cassidy, 2006). But now there are various popular websites which help to communicate and enlist profile like Friendster, MySpace, and Facebook (Ahmad, 2011). The Internet is full of these websites. Facebook page and YouTube channels too can be created. Followers can get all the information on one platform. Similar thinking people can come across on the same stage, and they can discuss the things and topics on that platform.

Social networking among youngsters is not just a fad, but it is part of their culture. A parent or a teacher cannot expect a teenager to abandon the internet or social networking. A youngster can thrive in modern society as computers, technology, the internet and social networking is utilized in their lifestyle. This is the reason why social networking should be used as a valuable tool and should be utilized as an opportunity to teach students (Jayme, 2011). The only solution of this is changing the way they use this platform. Training can be given to them, and a habit can be inculcated in students or teenagers so that they can use it in beneficiary manner.

Approximately, "85% of undergraduate students are Facebook users". These are only few, who are growing year by year, and they are only Facebook users, there are numerous of other social networking tools like YouTube, MySpace, WhatsApp (Stollak, Vandenberg, Burklund, & Weiss, 2011). Students like to join various communities on social networking like educational, entertainment, information and for any other use. 22% used Social Networking Sites for educational purposes, 38% for entertainment, 48% for Information and 18% for any other use (Mehmood & Taswir, 2013). It was found that 78.3% was spending the most time on Facebook, 58.4% on YouTube, 41.7% in reading blogs, 26.3% on Twitter, 29.2% of MySpace and 43% on Linked In (Stollak, Vandenberg, Burklund, & Weiss, 2011).

The problem is multitasking. It is also said, "…the problem is that most people have Facebook or other social networking sites, their e-mails and maybe instant messaging constantly running in the background while they are carrying out their tasks…" (Mehmood & Taswir, 2013). The University of New Hampshire agrees that the students grew up in the era of technology and social networking and now it is a part of their daily routine. 63% of heavy users received high grades, compared to 65% of light users.

For better learning, if social media is used, there are opportunities to collaborate, express, participate and reflect. It provides opportunities to help the teacher in feedback, knowledge sharing, notes sharing, knowledge storage and process and writing are also expanded (Crook, 2008). The boring environment at the school can be improved if social media gets involved as by the use of social media, they can expand the opportunities to contact the school and the students for this year's groups to the alumni and future students. It increases participation in activities at the school in addition to teaching. Also, it creates better contact with absent students and opportunities and to create virtual help in homework (McNely, 2009).

MAIN FOCUS OF THE CHAPTER

Major Objectives of the Chapter:

- To utilize the time wasted in social networking, chats, etc.
- To study the use of social networking learning.
- To brief upon the positive relationship between social media usage and students.
- To understand the role of social media used by teachers.

Research Methodology:

- Primary Data was collected with the experimental study in a school where age group of 10 to 17 years was involved, and a survey was conducted with a questionnaire and response was recorded.
- Secondary Data was collected from research papers, magazines, and online journals, etc.

Results:

1. In Table 1, the data of students using social networking websites by various classes is given. It is to be noted that the given school is in a metropolitan city. The users can vary if this survey was conducted in the village or some rural area and the result would have been different. In a rural area, the technology is not so easily available, and many other problems are also present.

The results show that around 86% of the students of the school from classes 5th and above use social media in their routine life.

2. A survey was also conducted to know how many students know the use of social networking as a learning aid. Table 2 displays the data on the number of students who use the social media to learn or to communicate on a topic related to the subject and curriculum. Students may have communicated to the teachers, parents or tutors related to a topic or curriculum using social media.

Table 1. Use of social media by various classes

Class	Total Number of Students	Number of Students Using Social Media	Percentage of Users
5	70	49	70.0
6	80	72	90.0
7	79	62	78.5
8	100	79	79.0
9	81	79	97.5
10	92	88	95.7
11	72	68	94.4
12	150	127	84.7
Total	**724**	**624**	**86.2**

Table 2. Use of social media by students of different classes for learning

Class	Total Number of Students	Social Media Usage for Learning	Percentage of Users
5	70	00	0.0
6	80	6	7.5
7	79	6	7.6
8	100	7	7.0
9	81	12	14.8
10	92	10	10.9
11	72	20	27.8
12	150	56	37.3
Total	**724**	**117**	**16.2**

As depicted in the table, it can be easily seen that only 16% of the students of the school from class 5th onwards use social media for the purpose of learning, which is very low.

Table 3 displays how many educators use social media as a teaching aid in the school to impart knowledge to the students.

It can be seen that only 21% teachers in the given school use social media as a teaching aid which is such a low figure in today's era of digitization.

3. Table 4 shows the behavior of different age group people that do they think before they post on social media or not.

Table 3. Use of social media by various teachers in teaching

Subject	Total Number of Teachers	Number of Social Media Users as Teaching Aid
Science	8	2
Mathematics	6	2
Language	12	1
Commerce	3	3
Others	37	6
Total	**66**	**14**

Table 4. Do people think before they post on social media or not?

Category	Always	Rarely	Never
Teachers	60%	30%	10%
Students of age above 15	20%	30%	50%
Students of age less than 15	12%	18%	70%

Clearly, it can be seen that even the teachers don't think what they are posting and forwarding on such social media platforms. So a training to handle the post and writing post is also required to a large extent.

Websites used by Teachers: In an attempt, teachers were asked to name the social networking websites used by them and what they professionally used to connect the parents and students. While only some teachers allow connecting themselves with the parents and students, the teachers connected to the students are better aware of student's problems and the way how to deal with it. Teachers connected with parents have better understanding between teacher-student and child-parent and parent-teacher relationship. Also, on a daily basis, parents can track their child's learning. The most used websites by teachers are -

1. **Twitter:** It keeps them up to date. Teaching is a profession where keeping yourself up to date is necessary. In today's world, it becomes the priority list whatever happening in the World must be part of the classroom the class should incorporate tell the relation of news update in the student's life. Even Students are also connected to other things which they can also discuss in the class. The teacher must have the knowledge of these topics live classroom should incorporate the news update about the topic. Teachers can tweet on Twitter about the next topic, social issue, and discuss for the change in syllabus.
2. **YouTube:** A video from the topic can be shown in the starting of the class that can be helpful for the class students to make, it more live. It makes students more attentive and creates interest in the topic. Many teachers use YouTube; some use reference of some movies, some use smart classes for the purpose. YouTube is a collection of videos. Teachers can find a video from it related to the topic

3. **Facebook:** Teachers have different opinions over the use of Facebook. Some teachers use it to become friend with the Student so that they can ask questions casually. While others have a point of view that student should respect teachers. They can have the formal talk but not on social networking, but the social networking uses informal talk which they don't like. Facebook provides a platform to share your views, topics, links, images which can help students to learn the topic if images link provided outside the class.

SOLUTIONS AND RECOMMENDATIONS

The time spent by students on the internet and specifically on social networking sites is a huge challenge for parents and teachers. Students not only waste their study time but often quietly stay online at late night, leaving them short of sleep and irritable behavior the next day when they come back to school. Over the past few years, the way of communication, the way of traveling and almost everything has been changing, but the education system chosen for educating used in the past is using in the present also. The lecture delivery method, the practical method, the examination system, etc. are all the same, and nothing has changed with time. Only a slight variation can be seen in the lecture delivery method that earlier books were used and now power point presentations are used for teaching. But there is hardly any interactive thing added in learning process over the period.

But this can be improved by training the teachers how they can utilize the resource of social networking as a teaching aid and feedback system. Social media's like, view and comment features can motivate the students too. Mostly teacher in the classroom does the one-way teaching. But teaching is not only imparting subject content but it is also about the real-life application of that concept. Along with subject content, the teacher should:

- Develop an understanding of concept and application of the concepts.
- Develop reasoning and thinking an ability.
- Inculcate decision-making an ability.
- Help in developing self-concept, and values.
- Develop tolerance and ambiguity, risk taking capacity, scientific temper, etc.

But there is no time for communication in class or taking problems of students in the class. Some students are of shy nature, and they are not able to present their view. The teacher can ask a question from one or two students, or if the teacher asks a question from the whole classroom also, only two or three students answer the question. Social media help the student to ask question after classroom time, and shy student can ask a question by sending a question and get their problem solved.

Rational thinking is required while speaking in public. It is required to think what will be the pros and cons of that particular post. It targets to which community and how to present the content. As public speaking is an art posting on social networking is also an art. They just post whatever they like. Even there is a forward facility in social media. Many of messages and posts are forwarded without reading and thinking. For example, advertisers target a particular age group, the particular community, so they create advertisement accordingly. Blog writing, critical thinking before writing, Facebook post writing on relevant content can be promoted. Presenting views on a particular topic socially can also be promoted. Moreover, if the content is of good quality, the number of likes, views, and comments increase which give them a sense of self-motivation. But to accomplish all this, training is required.

FUTURE RESEARCH DIRECTIONS

The customized and new way of using social media in teaching is required. More research on the teaching aid is required. A better way to involve the learner in teaching learning process is required in which social media can help. Training of teacher and students is required so that misuse of social media can be changed as a useful learning aid. The examination can also be minimized by feedback, review and can grade by comments, feedback, and discussion on that social media.

CONCLUSION

Students of all classes use social media in their daily routine. Students can use social media for learning either to make assignments or to share notes. Teachers, if trained appropriately, can use social media as a teaching aid by a discussion on Twitter, Facebook, and blogs on which students can present their views at the same time or after the class. Whenever or wherever they get stuck, they can post the problem. In this way, the learner can learn, and teacher can get feedback too. Consequently, it will make teaching learning process easier, convenient, interactive, lively and student friendly.

REFERENCES

Ahmad, A. (2011). Rising of Social Network Websites in India Overview. *IJCSNS International Journal of Computer Science and Network Security*, *11*, 156.

Cassidy, J. (2006). Me media: How hanging out on the Internet became big Business. *The new Yorker*, *82*, 13-50.

Crook, C. (2008). Web 2.0 technologies for learning: The current landscape – opportunities, challenges, and tensions. Nottingham: Becta: Leading next generation learning.

Jayme. (2011). The University of Colorado. Social Networking: The Unharnessed Educational Tool. *Undergraduate Research Journal*, *4*(1).

McNely, B.J. (2009). Backchannel Persistence and Collaborative Meaning. In *Proceedings of SIGDOC'09*, October 5 – 7.

Mehmood, S., & Taswir, T. (2013). The Effects of Social Networking Sites on the Academic Performance of Students in College of Applied Sciences. *International Journal of Arts and Commerce*, *2*, 113.

Stollak Matthew, J., Vandenberg, A., Burklund, A., & Weiss, S. (2011). Getting Social: The Impact of Social Networking Usage on Grades among College Students. *Proceedings of ASBBS*, *18*, 859–860.

KEY TERMS AND DEFINITIONS

Information and Communication Technology: It is a term used to communicate by using electronic media.

Social Learning Tools: It is a phrase that is used to reflect the use social networking functionality in the education setting.

Social Media: A platform that gives a common platform for sharing their news, views, and opinions regarding the happenings around them.

Social Networking Platform: A tool that is specially designed to communicate and discuss the issues with students, parents, and educators collaborate online.

Chapter 11
An Exploratory Study of Classroom and Online Teaching Practice in Relation to Inquisitiveness, Attitude, and Teaching Effectiveness

Neetu Singh
Dayalbagh Educational Institute (Deemed University), India

ABSTRACT

The present study is aimed at achieving main objectives i.e. to study the effect of Inquisitiveness and Attitude on Teaching Effectiveness among pupils' teachers practicing classroom-teaching practices; to study the effect of Inquisitiveness and Attitude on Teaching Effectiveness among pupils teachers practicing online-teaching practices, to explore the effectiveness of online teaching practice, to find out the future horizons of online teaching practice in Indian perspective. The sample of the present study is pupil teacher studying in Dayalbagh Educational Institute Deemed University Agra. 150 pupil teachers practicing classroom teaching and 150 pupil teachers practicing online teaching have been selected. T-Test, Linear Regression and SWOT Analysis have been used as statistical techniques. Pupil teachers possess more positive attitude and inquisitiveness towards online teaching practice in comparison to classroom teaching practice. Inquisitiveness and attitude significantly predicts the teaching effectiveness in online teaching practice and classroom teaching.

DOI: 10.4018/978-1-5225-3191-3.ch011

INTRODUCTION

Education is responsible for the upliftment of the human conditions. It is observed by Alade (2004) that the primary concern of education is the elevation of human conditions. Through education, people are enabled to develop their knowledge and skills, adopt new behavior and be able to survive in the society. A nation, though rich in national resources, cannot succeed unless its human resources are properly developed, and such a development is basically, a function of education. When a person communicates information or skill to another, it is described as the action of teaching, but not every way of bringing about learning in other counts is teaching, and not every act of teaching has a place within a program of education. Teaching is known as a complex art, and it depends on the method of teaching, curricula, infrastructure, technology, etc. Teaching is a series of actions by which a teacher makes efforts to bring desired behavioral changes in the students. The teacher attempts to change the manner of thinking, feeling and acting of students by their teaching.

Therefore, teaching is a program which helps the child to respond to his environment in an effective way. Teaching expands the compliance of students to complex environments. It communicates useful information to the students and develops a harmonious relationship among the teacher, the students, and the curriculum. Therefore, teaching plays a most important role in making students educated, learned and future oriented. At present technology has been developed to its height, it has also been entered the teaching-learning process and made the educational environment changed and fully developed. So there are a different type of changes have been occurred in teaching-learning process such as face-to-face teaching, Online teaching, Synchronous Class, Asynchronous Class, etc.

Therefore, B.Ed. students are trained to teach the students. They have to teach their students in all these types of classrooms in the present time. If pupil teachers are fully trained for their profession, they work effectively. In the present study, teaching effectiveness has been studied with correlating their inquisitiveness and attitude for teaching.

Classroom Teaching Practice

Before assigning pupil teachers for teaching practice to different schools, they are equipped with different teaching skills in teaching practice session. When pupil teachers practice their teaching skills and do their internship in a classroom environment, this type of teaching practice is called classroom teaching practice. Classroom-based teaching includes: face-to-face interaction, proper discussion, lectures and large group teaching, Tutorials, Seminars, Laboratory/Practical teaching, social interaction, etc. They are trained for maintaining classrooms and teach the students with managing teaching methods, techniques, approaches, evaluation techniques, etc.

Online Teaching Practice

The development of the Internet has provided new prospects for storing, sharing, discovering and interacting with information, including the development of online learning environments. At present, pupil teachers are trained for teaching not only in a classroom environment but also for the online environment. There is online teaching practice is conducted in Dayalbagh Educational Institute (Deemed University) India. Pupil teachers are teaching the students by using software “BIG BLUE BUTTON” where they get the opportunity to practice teaching with students and using advanced technology.

Inquisitiveness

Inquisitiveness is defined as psychological behavior which is related to eagerness for knowledge, curiosity for learning and readiness for intellectual development. Inquisitiveness for teaching is a basic form of interest in teaching. As the pupil teachers possess inquisitiveness for teaching, so they make efforts to join the profession of teaching and to make their teaching effective. They want to teach, instruct and interact with the students as well as explore and enrich their knowledge day by day. Their inquisitive mind always makes them active to achieve what they want to achieve. In the online environment, pupil teachers use a different type of technologies and e-resources for teaching. They are attracted by these advanced technologies. Therefore, their inquisitiveness increases and they want to do more new things in their teaching.

Attitude

According to Taneja (1991), attitude is ‘a predisposition to perceive, feel or behave towards specific objects in a particular manner.’ Good (1973) has defined attitude as ‘the predisposition or tendency to react specifically towards an object, situation or value; usually accompanied by feelings and emotions.’ The attitude of teachers towards teaching plays an important role in shaping the attitude of students towards the learning. The attitude of a teacher towards teaching may influence his actions in the classrooms, which becomes critical to student learning. A positive attitude towards teaching is helpful to teachers in maintaining harmonious relations with their students which are characterized by sympathetic understanding and affection. Teaching process requires intelligence, eagerness, practical skills and a sense of duty and integrity. Consequently, selecting the right persons for teaching profession is of chief importance for the development of the educational system. Only right persons can perform the right type of work. Proper selection and training of pupil teachers is now a commonly accepted idea and is considered to be necessary for improving teaching effectiveness. The Secondary Education Commission (1952-53) rightly mentioned 'the teacher training program can engender the knowledge, skills, and attitudes which will enable the teacher to begin his task with a reasonable degree of confidence and with the minimum amount of experience.' Therefore, it is found necessary that the attitude of future teachers should be measured so that in cases where a negative attitude towards teaching has been found could be so modified that a positive attitude towards teaching profession can be cultivated in them.

Teaching Effectiveness

Teaching is a dynamic and complex phenomenon, including teachers, students and subject matter. Effective teaching cannot be defined exactly because the criteria differ for every instructional situation and every teacher. Effective teaching leads to engaged and intelligent learning. It may be defined as helping students to learn how to do somewhat, providing with knowledge, besides beginning to know and understand. It is also about guiding and facilitating learning, enabling the learners to learn and arrange the conditions for learning.

According to Anderson (2004), there are various characteristics associated with an effective teacher such as Commitment, Confidence, Trustworthiness, Respect, Analytical Thinking, Conceptual thinking, Drive for improvement, Information seeking, Initiative, Flexibility, Accountability, Passion for learning.

According to Dickson (1980), "...teaching effectiveness is a demonstrated repertoire of competencies involved with teaching plans and materials, classroom procedure, inter-personal skills, and the learner's reinforcement and involvement reflected in the teacher behavior." Teaching effectiveness is concerned with what a teacher is and what he does in the classroom in order to encourage the pupil's learning. Therefore, teaching effectiveness includes such representative qualities of a teacher as he exhibits in the classroom with the intention that his students learn what he teaches, such actions of a teacher which he performs in the classroom as have the intention that his students learned what he teaches.

BACKGROUND

LaPrade, Gilpatrick and Perkins (2014) assessed the effective online instructional practices in higher education at Grand Canyon University/GCU. The GCU College of Education created a formative evaluation tool for online faculty—Five by Five—based on present recommended best practices in online teaching. The tool has been used by online faculty to reflect on their performance using the five criteria i.e. communication, engagement, expertise, and use of quality instruction techniques. A quasi-experimental research was conducted with a sample of fifty online instructors. The sample was divided equally into two groups experimental and control groups. The sample selected in the experimental group was assigned to do a self-evaluation and reflection on their online teaching capabilities using the Five by Five tools. The results revealed statistically significant positive effects of the utilization of Five by Five by online faculty in terms of their participation in the discussion forums and their use of quality instructional techniques.

Scagnoli, Buki and Johnson (2007) studied about the integration of online technologies in educational practice is rendering new opportunities for teaching and learning process. It is a fact that teachers who have taught fully online courses have acquired new skills and have had the opportunity to implement novel pedagogical practices in the online environment. Though, it is unclear whether direct exposure to fully online teaching facilitates the integration of technology in traditional classrooms. This qualitative investigation studied the transfer of four experienced faculty members' pedagogical practices from online to face-to-face teaching. This case study showed that the instructors' online teaching experience influences their perceptions and understanding of online pedagogical strategies, and the transmission of pedagogical strategies back to the classroom is a complex process influenced by the instructors' teaching style, satisfaction with working in the online environment, and the similarity of content and context between online and face-to-face courses.

Ya Ni (2013) explored that public administration programs extend online education offerings to reach more time and place-bound students, and as accredited institutions become interested in documenting teaching and learning effectiveness, the degree to which online students are successful as compared to their classroom counterparts is of interest to teaching faculty and others charged with assessment. By comparing student performance measures and assessments of learning experience from both online and traditional sections of a required graduate public administration research methods course taught by the same teacher, this paper provides evidence that student performance as measured by grade is independent of the mode of instruction. Persistence in an online environment may be more challenging in research methods classes than in other public administration classes.

Omara, Hassana & Atan (2012) aimed their research to identify learner's attitudes toward e-mentoring. A survey analysis was conducted. A total of 205 participants were selected for collecting the data. The results presented that learner's attitudes play a significant role in predicting e-mentoring.

Vijayalakshmi (2005) carried out research to study the attitude towards Educational Technology among B.Ed. students of Bharathidasan University. This research concluded that B.Ed. students have a favourite attitude towards Educational Technology.

A study by Gopal and Anandan (2013) indicates that the B.Ed. Students are having a lesser attitude towards e-learning for the Classroom Instruction, there was no significant difference between the mean scores of the attitude of B.Ed. students towards e-learning for classroom instruction with respect to Gender and Parental qualification and there was a significant difference between the scores of the attitude of B.Ed. students towards e-learning for classroom instruction with respect to their discipline of the Subject wise. It is concluded that the B.Ed. Students having more access over the Internet browsing abilities.

A research paper was written by Victor (2013). This research paper seeks to explore one of the important dimensions of teacher trainees: Attitude towards Information and Communication Technology (ICT). The research focuses on the level of attitude towards ICT among B. Ed teacher trainees. The sample was selected of 952 teacher trainees. The findings indicated that majority of teacher trainees (60.5%) showed uncertainty in their attitude and only 39.5% of the trainees showed positive attitude towards ICT. The study recommends that crash courses especially in ICT applications and Workshops on Modern ICT tools could be conducted to make the teacher trainees aware of the different innovative techniques and methods, which will enhance the attitude towards ICT.

Toor (2014) studied teacher effectiveness, general intelligence, and creativity of secondary school teachers in relation to the type of school. The result of the study revealed that there is a significant difference in the teacher effectiveness of government and private secondary school teachers. The government school teachers are more effective than private school teachers.

Sharadha and Paremeswaram (2008) made an attempt to study the role of some behavioral variations among teachers and their possible implications for effective classroom teaching and learning. The results of the study showed that the management of the institutions had a significant impact on the behavior variations in the level of their teaching effectiveness.

Vijayalakshmi (2005) aimed to find out the effect of locality, management, and subject of teaching on teacher effectiveness and job satisfaction. The findings of the study revealed that management of school has a significant impact on both teacher effectiveness and job satisfaction.

After reviewing previous studies conducted on online teaching environment, inquisitiveness, attitude and teaching effectiveness, it has been found that no study has been conducted to explore whether inquisitiveness and attitude of pupil teachers influence their teaching effectiveness. Therefore, the present study has been designed by the researcher.

MAIN FOCUS OF THE CHAPTER

Objectives

1. To compare the Inquisitiveness among the pupil teachers practicing classroom and online teaching practice.
2. To compare the pupil teachers' attitude practicing classroom and online teaching practice.
3. To compare the Teaching Effectiveness among the pupil teachers practicing classroom and online teaching practice.
4. To study the effect of Inquisitiveness on Teaching Effectiveness among pupil teachers practicing classroom teaching practice.
5. To study the effect of Attitude on Teaching Effectiveness among pupil teachers practicing classroom teaching practice.
6. To study the effect of Inquisitiveness on Teaching Effectiveness among pupil teachers practicing online teaching practice.
7. To study the effect of Attitude on Teaching Effectiveness among pupil teachers practicing online teaching practice.
8. To explore the effectiveness of online teaching practice.
9. To find out the future horizons of online teaching practice in Indian perspective.

Hypothesis

1. There exists no significant different in Inquisitiveness among pupil teachers practicing classroom and online teaching practice.
2. There exists no significant different in attitude among pupil teachers practicing classroom and online teaching practice.

3. There exists no significant different in Teaching Effectiveness among the pupil teachers practicing classroom and online teaching practice.
4. There exists no significant effect of Inquisitiveness on Teaching Effectiveness among pupil teachers practicing classroom teaching practice.
5. There exists no significant effect of Attitude on Teaching Effectiveness among pupil teachers practicing classroom teaching practice.
6. There exists no significant effect of Inquisitiveness on Teaching Effectiveness among pupil teachers practicing online teaching practice.
7. There exists no significant effect of Attitude on Teaching Effectiveness among pupil teachers practicing online teaching practice.

Research Methodology

This study follows quantitative as well as qualitative methodology of research -

- **Variables:** There are three variables considered in the present study out of which two variables are independent and one variable is the dependent variable. Table 1 shows the variables of the present study.
- **Research Method:** Descriptive survey research method has been employed by the researcher to achieve the objectives of the present study.
- **Sampling Procedure:** Sample of the present study is pupil teachers studying in B. Ed. course in Faculty of Education, Dayalbagh Educational Institute (Deemed University) Agra. There are 300 pupil teachers in Faculty of Education who were practicing teaching in the third semester. 150 pupil teachers are practicing classroom teaching, and 150 pupil teachers are practicing online teaching.
- **Research Instruments:** Standardized as well as self-developed questionnaires have been used to collect the information regarding the variables inquisitiveness, attitude and teaching effectiveness.

The description of the administered tools is given below:

Table 1. The variables of the present study

Dependent Variable	Independent Variable
Teaching Effectiveness	Inquisitiveness
	Attitude

- **Inquisitiveness Inventory:** The researchers have developed an inquisitiveness inventory to measure the inquisitiveness for teaching among the pupil teachers. This inventory consists of total 26 items, and all these items of Inquisitiveness Inventory are positively worded. This inventory is a five-point scale and contains a score of 5,4,3,2, and 1 for 'strongly agree,' 'agree,' 'undecided,' 'disagree' and 'strongly disagree.' The Cronbach's Alpha Reliability for this scale is 0.75, and content validity is 0.81.
- **Teacher Attitude Inventory:** Teacher Attitude Inventory by Dr. S. P. Ahluwalia is a Likert instrument which consists of six dimensions, i.e., Teaching Profession, Classroom Teaching, Child-Centred Practices, Educational Process, Teachers and Pupils. Each dimension has 15 items pertaining to a particular aspect of prospective and practicing teachers. Therefore, this inventory comprises of total 90 items including both positive and negative items. The scoring for this inventory is given in table 2.

Reliability of this inventory was estimated by Split-half (odd even method) method and found to range from 0.79 corrected up to (0.88) to 0.54 (Rational Equivalence KR 21). The test-retest reliability coefficients were found to be 0.58 and 0.64 after an interval of 3 months and 9 months. The teacher attitude inventory seems to have content validity compared with the scores of MTAI developed by Dr. M. C. Joshi. The obtained correlation coefficients for prospective teachers were positive. For the dimensions, Teaching Profession, Classroom Teaching, Child-Centred Practices, Educational Process, Teachers, Pupils correlation coefficients were + 0.42, + 0.32, + 0.02, + 0.27, + 0.23, + 0.07 respectively and it is found +0.23 for the total.

- **Teacher Effectiveness Scale:** Teacher Effectiveness Scale by Pramod Kumar and D. N. Mutha (1999) is a Likert type scale. This scale presents teachers' effectiveness in six areas, i.e., academic, professional, social, emotional, moral and personality. This is a five-point scale with alternatives of 'strongly agree,' 'agree,' 'undecided,' 'disagree' and 'strongly disagree' with scores

Table 2. Scoring for Teacher Attitude Inventory

Scale	Positive	Negative
Strongly Agree (SA)	4	0
Agree (A)	3	1
Undecided (U)	2	2
Disagree (D)	1	3
Strongly Disagree (SD)	0	4

of 5, 4, 3, 2 and 1. There are total 69 items in this scale and all the items are positively formulated. The Split-Half Reliability for this tool is 0.67 with an index of reliability of 0.82. The test-retest reliability of this scale is 0.75 with an index of reliability of 0.85 with two-month interval gap. This scale has been validated against principal's rating. The correlation between principal's rating and self-rating is 0.70 with an index of reliability of 0.87.

- **Statistical Techniques:** Various statistical techniques have been used by the researcher to analyze the data. T-Test, Linear Regression, and SWOT Analysis have been employed in this study.

SOLUTIONS AND RECOMMENDATIONS

Objective 1: To Compare the Inquisitiveness Among the Pupil Teachers Practicing Classroom and Online Teaching Practice

T-test has been used to find out the difference in the inquisitiveness among the pupil teachers practicing classroom and online teaching practice. This difference is shown in Table 3.

This table presents that mean values of inquisitiveness for classroom teaching, and online teaching is 81.93 and 103.40 respectively. The obtained t-value is 4.676 which is significant at 0.05 levels. Therefore the null hypothesis that "There exists no significant different in Inquisitiveness among pupil teachers practicing classroom and online teaching practice" is rejected. It means that significant difference is found in inquisitiveness among pupil teachers practicing classroom teaching practice and online teaching practice. The reason behind it may be that present generation normally uses mobiles, Androids, laptops, social networking sites, etc. They are more comfortable with ICT and aware for digital literacy. They feel boredom due to the use of lecture method and no innovations in the traditional way of teaching. In classroom teaching, pupil teachers prepare lesson plans and teach the students by

Table 3. Exhibiting difference in the inquisitiveness among the pupil teachers practicing classroom and online teaching practice

Different Teaching	N	Mean	SD	t-value	Level of Significance
Classroom Teaching Practice	150	81.93	6.75	4.676	0.05
Online Teaching Practice	150	103.40	11.72		

using some traditional examples and teaching aids. While pupil teachers are curious to learn how to use computers and e-resources for teaching purpose. They like to do innovations in their teaching and presentations as well as they want to make their teaching more interesting. Therefore, it can be said that the pupil teachers practicing online teaching are more inquisitive in comparison to the pupil teachers practicing classroom teaching.

Objective 2: To Compare the Pupil Teachers' Attitude Practicing Classroom and Online Teaching Practice

T-test has been used to find out the difference in attitude among the pupil teachers practicing classroom and online teaching practice. This difference is shown in Table 4.

This table shows that mean values of attitude for classroom teaching and online teaching is 194.91 and 225.53 respectively. The extent of t-value is found 3.13 which is significant at 0.05 levels. Therefore, the null hypothesis that "There exists no significant different in attitude among pupil teachers practicing classroom and online teaching practice" is rejected. It means that significant difference is found in attitude among pupil teachers practicing classroom teaching practice and online teaching practice. It is found that the pupil teachers practicing in online teaching have more positive attitude in comparison to the pupil teachers who were practicing in classroom teaching. It may be due to the use of online resources in teaching. The pupil teachers practice in a techno-centric environment of teaching-learning; they are more techno-friendly; they feel more active and learn something new continuously as well as generate the content by their own efforts. All these things develop positive thinking for the teaching profession. On the other hand, pupil teachers practicing in classroom teaching are provided the traditional ways of teaching practice. They take it as their assignment and think about it to just complete the assignment. Therefore, pupil teachers practicing in online teaching possess more positive attitude than pupil teachers practicing in classroom teaching.

Table 4. Exhibiting difference in the attitude among the pupil teachers practicing classroom and online teaching practice

Different Teaching	N	Mean	SD	t-value	Level of Significance
Classroom Teaching Practice	150	194.91	23.779	3.13	0.05
Online Teaching Practice	150	225.53	29.22		

Objective 3: To Compare the Teaching Effectiveness Among the Pupil Teachers Practicing Classroom and Online Teaching Practice

T-test has been used to find out the difference in Teaching Effectiveness among the pupil teachers practicing classroom and online teaching practice. This difference is shown in Table 5.

This table shows that mean values of Teaching Effectiveness for classroom teaching and online teaching is 196.66 and 228.73 respectively. The calculated t-value is found 3.05 which is significant at 0.05 levels. Therefore, the null hypothesis that "There exists no significant different in Teaching Effectiveness among the pupil teachers practicing classroom and online teaching practice" is rejected. It means that significant difference is found in Teaching Effectiveness among pupil teachers practicing classroom teaching practice and online teaching practice. It is found that the pupil teachers practicing in online teaching teach their students effectively in comparison to the pupil teachers who were practicing in classroom teaching. Pupil teachers take an interest in online teaching practice, they prepare attractive and effective presentations to teach and try to give real life examples related to the content, and they make their teaching effective by using audio-visual aids. On the other side, they lack all these things in classroom settings. Therefore, teachers effectively teach the students in an online environment.

Objective 4: To Study the Effect of Inquisitiveness on Teacher Effectiveness Among Pupil Teachers Practicing Classroom Teaching Practice

Linear Regression has been used to determine the effect of Inquisitiveness on Teacher Effectiveness among pupil teachers practicing classroom teaching practice. This effect is visualized in Table 6.

This table indicates the coefficient of multiple determination R^2 0.436 and F-value is 114.61 which is significant at 0.05 level. This means that 43.6% of the variation

Table 5. Exhibiting difference in teaching effectiveness among the pupil teachers practicing classroom and online teaching practice

Different Teaching	N	Mean	SD	t-value	Level of Significance
Classroom Teaching Practice	150	196.66	34.96	3.05	0.05
Online Teaching Practice	150	228.73	48.81		

Table 6. Exhibiting the effect of inquisitiveness (predictors / constant) on teacher effectiveness (dependent variable) among pupil teachers practicing classroom teaching practice

Model	R	R Square	Adjusted R Square	Std. Error of the Estimate	F	Level of Significance
1	0.661	0.436	0.433	17.084	114.61	0.05

in teaching effectiveness of pupil teachers (who were practicing classroom teaching practice) can be significantly accounted by their inquisitiveness.

Table 7 illuminates that at 0.05 level of significance, the inquisitiveness of the students significantly predicts the teaching effectiveness of the pupil teachers. In the case of inquisitiveness, positive regression weights (1.44) are found which shows that the independent variable 'inquisitiveness' positively influences the teaching effectiveness.

Objective 5: To Study the Effect of Attitude on Teaching Effectiveness Among Pupils Teachers Practicing Classroom Teaching Practice

Linear Regression has been used to determine the effect of Attitude on Teaching Effectiveness among pupil teachers practicing classroom teaching practice. This effect can be seen in Table 8.

This table presents the coefficient of multiple determination R^2 0.386 and F-value is 93.03 which is significant at 0.05 level. This means that 38.6% of the variation in

Table 7. Exhibiting coefficients

Model		Unstandardized Coefficients		t	Level of Significance
		B	Std. Error		
1	(Constant)	122.16	14.326	8.52	0.05
	Inquisitiveness	1.44	0.135	10.70	0.05

Table 8. Exhibiting the effect of attitude (predictors/constant) on teaching effectiveness (dependent variable) among pupil teachers practicing classroom teaching practice

Model	R	R Square	Adjusted R Square	Std. Error of the Estimate	F	Level of Significance
1	0.621	0.386	0.382	17.51	93.03	0.05

teaching effectiveness of pupil teachers (who were practicing classroom teaching practice) can be significantly explained by their Attitude.

Table 9 exemplifies that at 0.05 level of significance, Attitude significantly predicts the teaching effectiveness of the pupil teachers. The positive regression weight (0.586) shows that the independent variable 'Attitude' positively influences the teaching effectiveness.

Objective 6: To Study the Effect of Inquisitiveness on Teacher Effectiveness Among Pupil Teachers Practicing Online Teaching Practice

Linear Regression has been used to identify the effect of Inquisitiveness on Teacher Effectiveness among pupil teachers practicing online teaching practice. This effect is visualized in Table 10.

This table shows the coefficient of multiple determination R^2 0.707 and F-value is 357.88 which is significant at 0.05 level. This means that 70.7% of the variation in teaching effectiveness of pupil teachers (who were practicing online teaching practice) can be significantly accounted by their inquisitiveness.

Table 11 elucidates that at 0.05 level of significance, Inquisitiveness significantly predicts the teaching effectiveness of the pupil teachers practicing online teaching practice. The positive regression weight (2.29) shows that the independent variable 'Inquisitiveness' positively affects the teaching effectiveness.

Table 9. Exhibiting coefficients

Model		Unstandardized Coefficients		Standardized Coefficients	t	Level of Significance
		B	Std. Error	Beta		
1	(Constant)	138.40	14.721	0.621	9.40	0.05
	Attitude	0.586	0.061		9.64	0.05

Table 10. Exhibiting the effect of inquisitiveness (predictors / constant) on teacher effectiveness (dependent variable) among pupil teachers practicing online teaching practice

Model	R	R Square	Adjusted R Square	Std. Error of the Estimate	F	Level of Significance
1	0.841	0.707	0.705	10.014	357.88	0.05

Table 11. Exhibiting coefficients

Model		Unstandardized Coefficients		t	Level of Significance
		B	Std. Error		
1	(Constant)	32.19	9.98	3.22	0.05
	Inquisitiveness	2.29	.122	18.91	0.05

Objective 7: To Study the Effect of Attitude on Teacher Effectiveness Among Pupil Teachers Practicing Online Teaching Practice

Linear Regression has been used to identify the effect of Attitude on Teacher Effectiveness among pupil teachers practicing online teaching practice. This effect can be seen in Table 12.

This table shows the coefficient of multiple determination R^2 0.564 and F-value is 191.72 which is significant at 0.05 level. This means that 56.4% of the variation in teaching effectiveness of pupil teachers (who were practicing online teaching practice) can be significantly accounted by their Attitude. Table 13 explicates that at 0.05 level of significance, Attitude significantly predicts the teaching effectiveness of the pupil teachers practicing online teaching practice.

The positive regression weight (0.955) shows that the independent variable 'Attitude' positively affects the teaching effectiveness.

Table 12. Exhibiting the effect of attitude (predictors/constant) on teacher effectiveness (dependent variable) among pupil teachers practicing online teaching practice

Model	R	R Square	Adjusted R Square	Std. Error of the Estimate	F	Level of Significance
1	0.75	0.564	0.561	12.22	191.72	0.05

Table 13. Exhibiting coefficients

Model		Unstandardized Coefficients		t	Level of Significance
		B	Std. Error		
1	(Constant)	53.30	12.11	4.39	0.05
	Attitude	0.955	.069	13.84	0.05

Objective 8: To Explore the Effectiveness of Online Teaching Practice

SWOT Analysis has been conducted in the present study to explore the effectiveness of online teaching-learning environment. A SWOT analysis presents the positive and negative aspects of the organization (i.e. Strength & Weakness) and outside of it, in the external environment (Opportunity & Threat).The researcher has collected information and conducted SWOT analysis after taking interviews of 50 pupil teachers. Therefore the strengths, weaknesses, opportunities and threats of online teaching-learning environment are as follows -

STRENGTHS

Strength of online teaching environment has been given from pupil teachers and students' perspective which is discussed below -

Pupil Teachers' Perspective

- Pupil teachers are in the learning phase in which they are learning how to teach, and use of technology is creating interest among them for teaching.
- Online teaching-learning environment is making teaching easy and effective because of using audio-visual presentations as these teaching aids are providing real life examples of the content learning.
- Pupil teachers generate the content by their efforts via using content generation tools and upload the content on LMS (Learning Management System).
- Pupil teachers access more information and make the content enrich for students.
- Pupil Teachers use different online tools to evaluate the progress of the students. They evaluate students' ability in major fields of their learning ability i.e. their basic knowledge, understanding, practical knowledge, creativity, etc.
- The online teaching makes teaching-learning process more enjoyable. Use of advanced technology and online presentations encourages pupil teachers to add more in their teaching and make the class effective and more interactive.
- Online learning makes collaboration more effective by organizing group projects. Online tools, apps, and programs offer a unique setting for pupil teachers to engage in a group project.

- Teaching practice in an online environment is motivating. The teaching and content generation related tasks, immediate feedback, formative assessment motivate and help in increasing self-confidence, and progressively makes them express their involvement in teaching-learning, and take responsibility.

Students' Perspective

- Multimedia presentations encourage debate; existing technical means allow recreation of processes and their simulation. Image, dynamic, sound and word are stimulating the connectivity regarding: cognitive processes (sensory and rational, analytic and synthetic, the representation of thinking) content, people who interact.
- Students learn at their pace by recording the lessons and revise it whenever required.
- The online presentations are more interesting, so students are taking an interest in learning.
- It creates curiosity for learning among the students, they like more to use online apps for learning and completing the tasks and projects.
- Online learning prepares students for their future and to make them techno-centric.
- It reduces the boredom of classroom environment and presents various other ways of teaching with the help of online apps and tools.
- Students like to participate in online interaction with teachers in comparison to classroom interaction. The hesitation is also decreased in online teacher-student's interaction because they can also ask questions to the teachers and also respond only by typing if they do not want to participate in the verbal interaction.

WEAKNESSES

- There is not sufficient advanced infrastructure including physical, financial, and human resources; inadequate capital funds to support.
- All teaching staff is not aware and not wants to learn to use the technology so that pupil teachers can be trained properly.
- The students who require special attention of the teachers, this way of teaching do not give them more opportunity to learn.

- A certain kind of reduction of the relations between learners, between them and the teacher (Ozuorcun and Tabak, 2012), a possible loss of direct communication and immediate collaboration once the call for learning technologies, the possibility of, by using e-learning to create dependence on technology and isolation of the learner, rather than amplifying interactions with those involved in the process (Canadian Council on Learning, 2009).
- Due to higher costs of online learning services, pupil teachers face difficulty for working at home.

OPPORTUNITIES

- A National Learning Management System can be developed for teaching practice.
- Online teaching provides flexibility in terms of timing, especially when it becomes difficult to accommodate certain class schedules.
- Online learning increases interest for different categories of beneficiaries for e-learning educational services, various researches show that the interest in the arrangement of e-learning systems is growing higher, and it is visible.
- It meets radical transformation of all aspects of education (from access to obtaining diplomas, from final to results, from process to infrastructure, from teaching to evaluation, from teacher to students) as a result of technology dynamics (Demiray, 2010; Motschnig-Pitrik & Standl, 2012).
- Online teaching practice encourages digital learning among the pupil teachers, so that they also can make their students learn to use the online resources.
- Expansion of technology causes the latter to become a perfect environment for expression and development of e-learning educational services (Clark & Mayer, 2008, 2011; Motschnig-Pitrik & Standl, 2012; Liebowitz & Frank, 2011; Welsh, Wanberg, Brown, & Simmering, 2003; Dobre, 2010).

THREATS

- There are many students who belong to the economically weak background, and they do not have money for purchasing computers and using e-resources. There are many students who cannot afford a computer are at a real disadvantage and are forced to fight with crowds in a noisy computer lab.
- Knowledge of high-tech teaching methods is now a requirement for professors.
- Classroom learning is better than online teaching for students because it helps in teaching the socialization.
- It encourages organizational change e.g. departmental restructuring, cost cuttings, etc.

- The absence of face to face on-going monitoring, the inadequate degree of development of their responsibility regulates the abandonment of consistent learning efforts.
- It also generates negative effects such as: the danger of ignoring the student (Clark & Mayer, 2008, 2011), the possibility of producing an entire generation "of noncritical thinkers" (Liebowitz & Frank, 2011); technical problems (Demiray, 2010); lack of adequate infrastructure (lack of proper internet connection and very low bandwidth); "some difficulties of online administering: assuring the security of users, monitoring the students and offered services" (Dobre, 2010).
- Expenditures on e-learning educational services are not as small as they seem. There are broad categories of expenditures, such as expenditures on new technology (covering the degree of novelty incorporated into services, transmission of information in the network, maintenance of equipment, production of materials) (Dobre, 2010). New services require considerable investment in technology and human resources training, specific costs for designing and developing e-courses and achieving technology that allows guest to use the program (Welsh et al., 2003).
- There is a lack of quality controls and lack of a set of standards for e-content production and their delivery mechanisms.
- There are other threats, i.e.; long time is required to create and maintain e-learning courses (Arabasz et al., 2003), costs of training to update teaching methods and increased confidence in the new technologies.

Objective 9: To Find Out the Future Horizons of Online Teaching Practice in Indian Perspective

The researcher has also given future horizons of online teaching practice in Indian perspective. India is a developing country, and at present by following various effective governmental plans, policies and technological development, it is trying to come under the category of developed country. When its educational system is discussed, it is found that there are more efforts needed in this field to make literacy rates increase and technological interference is must extend the area of education and literacy. Therefore, if online teaching practice is conducted, the people who belong to the remote backward area will be able to receive an education. Online teaching and technology provide the opportunity to the teachers to create more interesting content and different effective ways of presenting the content before the learners so that the interest and curiosity can be developed among them. Today various types of teaching-learning are organized with the help of ICT like smart classes, m-learning, distance learning, virtual learning, blended learning, flipped classroom, etc. as the main vision of India is to increase the literacy rates and educational quality, all these plays a significant role.

There are various future horizons given by the researcher which can be followed to create e-environment of teaching-learning practice -

- Pupil teachers must be provided online teaching practice with classroom teaching practice so that they can use this in their future profession.
- Teacher educators should implement advance ways of teaching practice for the internship of pupil teachers.
- Literacy rates will be increased through the use of ICT.
- Students will be more learned when they learn with the use of more interesting content.
- Audio-visual presentations attract the learners and also create inquisitiveness to learn.
- Science and technological aspect of the country will be developed more.

After that, some suggestions have also given by the researcher for making the teacher training effective and successful in online environment -

- Adequate advanced infrastructure including physical, financial, and human resources must be provided with sufficient capital funds.
- Awareness programs to use the ICT must be conducted for making aware the teacher educators to learn ICT so that they can properly train the pupil teachers.
- It should be tried to train the pupil teachers how to deal the students who require special attention of the teachers in online teaching-learning environment.
- Pupil teachers must be familiarized with the high-tech teaching methods which are effectively used in online teaching.
- Pupil teachers should be learned about the importance of face to face interaction because it is must between teacher and students.
- Online teaching must be conducted by following few things i.e. aimed to develop critical thinking among the students, and none should be ignored in online class.
- Clear policies should be developed be the institutions and government also for quality controls and standards for e-content production and their delivery mechanisms.
- One more importance things that confidence must be increased in the new technologies among the pupil teachers.

Therefore, these suggestions can improve the online teaching-learning environment, confidence and interest can be developed for using ICT among teacher educators and pupil teachers and an advanced and modern method of teaching the students can be learned.

CONCLUSION

The positive attitude of pupil teachers towards their profession defines the parameters of teaching and learning. Therefore, the feelings, interests of teachers affect the performance of teachers. Teachers with positive attitude contribute more competently to educate the child. Effective teaching opens doors for students' learning and makes them free for an academic environment. In classroom environment of teaching practice, the pupil teachers have less positive attitudes and inquisitiveness for teaching in comparison to the pupil teachers who practice in online teaching. The reason of it may be that online learning environment is now a friendly platform for today's' generation. They enjoy when using technologies in teaching, due to it their inquisitiveness increase. Pupil teachers know about recent trends in education, so because of the being aware of entrance of technology in teaching-learning process, due to its demands and prospects their attitude is being positive. Positive attitude in teaching enables the teachers to create a learning community where every student has access to meaningful learning opportunities.

REFERENCES

Alade, I. A. (2004). Improving the status of vocational technical education: A route to National development. In D. F. Elaturoti & K. Babarinde (Eds.), *Teachers' Mandate on Education and Social Development in Nigeria* (pp. 95–105). Ibadan: Sterling-Horden Publishers.

Anderson, L. W. (2004). *Increasing teacher effectiveness* (2nd ed.). UNESCO: International Institute for Planning.

Arabasz, P., Pirani, J., Pond, S., & Fawcett, D. (2003). *Supporting E-Learning in Higher Education*. Retrieved from http://net.educause.edu/ir/library/pdf/ers0303/rs/ers0303w.pdf

Canadian Council on Learning. (2009). *State of e-learning and the science of instruction, proven guidelines for consumers and designers of multimedia learning* (3rd ed.). San Francisco: Pfeiffer.

Clark, R. C., & Mayer, R. E. (2011). E-Learning and the Science of Instruction, Proven Guidelines for Consumers and Designers of Multimedia Learning (3rd ed.). Pfeiffer: San Francisco.

Demiray, U. (2010). *E-Learning practices, Cases on challenges facing e-learning and national development: Institutional Studies and Practices (Vol. 1)*. Eskisehir, Turkey: Anadolu University.

Dickson, S. (1980). Personality and the Effective Teacher. *Journal of Teacher Education, 12*(3), 335–337.

Dobre, I. (2010). Studiu critic al actualelorsisteme de e-learning, Academia Română. Institutul de cercetăripentruinteligenţăartificială: Bucureşti.

Good, C. V. (1973). *Dictionary of Education*. New York: McGraw-Hill Book Company.

Gopal, B. V., & Anandan, K. (2013). Attitude towards e-Learning in Classroom Instruction among the B. Ed. Students at Colleges of Education. *Seminar on Current perspectives on Education*. Retrieved from http://www.languageinindia.com/feb2013/gopalanandan.pdf

LaPrade, K., Gilpatrick, M., & Perkins, D. (2014). Impact of Reflective Practice on Online Teaching Performance in Higher Education. *MERLOT Journal of Online Learning and Teaching, 10*(4). Retrieved from http://jolt.merlot.org/vol10no4/Gilpatrick_1214.pdf

Liebowitz, J., & Frank, M. (2011). *Knowledge Management and E-Learning*. U.S.A.: Taylor & Francis Group.

Motschnig-Pitrik, R., & Standl, B. (2012). *Person-centered technology enhanced learning: Dimensions of added value. Computers in Human Behaviour*. doi:10.1016/j.chb

Omara, N. D., Hassana, H., & Atan, H. (2012). Student Engagement in Online Learning: Learners Attitude toward E-Mentoring. *Procedia-Social and Behavioural Sciences, 67*. doi:10.1016/j.sbspro.2012.11.351

Ozuorcun, N. C., & Tabak, F. (2012). Is m-learning versus e-learning or are they supporting each other? *Procedia: Social and Behavioral Sciences, 46*, 299–305. doi:10.1016/j.sbspro.2012.05.110

Scagnoli, N. I., Buki, L. P., & Johnson, S. D. (2007). The Influence of Online Teaching on Face-To-Face Teaching Practices. *Journal of Asynchronous Learning Networks, 13*(2). Retrieved from http://files.eric.ed.gov/fulltext/EJ862351.pdf

Sharadha, D.G., & Parameswaram, E.G. (2008). Teacher Characteristics and Learning in the Classroom. *Edutracks*, *11*(3).

Taneja, R. P. (1991). *Dictionary of Education*. New Delhi: Anmol Publications.

Toor, K. K. (2014). A Study of Teacher Effectiveness, General Intelligence and Creativity of Secondary School Teachers. *MIER Journal of Educational Studies, Trends & Practices*, *4*(1).

Victor, S. R. (2013). Teacher-Trainees Attitude towards ICT. *Journal of Education and Practice*, *4*(19).

Vijayalakshmi, A. (2005). Teacher Effectiveness and Job Satisfaction of Women Teachers. *Edutracks*, *4*(7).

Welsh, E. T., Wanberg, C. R., Brown, K. G., & Simmering, M. J. (2003). E-learning: Emerging uses, empirical results and future directions. *International Journal of Training and Development*, *7*(4), 245–258. doi:10.1046/j.1360-3736.2003.00184.x

Ya Ni, A. (2013). Comparing the Effectiveness of Classroom and Online Learning: Teaching Research Methods. *Journal of Public Affairs Education*, *19*(2), 199–215. Retrieved from http://www.naspaa.org/JPAEMessenger/Article/VOL19-2/03_Ni.pdf

KEY TERMS AND DEFINITIONS

Attitude: Attitude can be defined as a psychological tendency that is expressed by evaluating a particular entity with some degree of favor or disfavor. In the present study, pupil teachers' attitude towards the teaching process has been studied.

Child-Centered Practices: Learner-centered education and student-centered learning are the synonyms of Child-centered practices. Child centered practices broadly encompass the teaching methods, techniques, teaching approaches, evaluation techniques and the whole teaching process which are used by teachers according to the mental level, interest, age, needs, IQ, etc. of the students.

Classroom Teaching Practice: When the student teachers practice their teaching activities in classroom environment and before the school students, this practice teaching is called classroom teaching practice.

Information and Communication Technology: Information and Communication Technology refers to technologies that provide access to information through telecommunications. It is similar to Information Technology but focuses primarily on communication technologies. This includes the Internet, wireless networks, cell phones, and other communication mediums.

Inquisitiveness: Inquisitiveness refers to the psychological behavior which is associated with the eagerness for knowledge, intellectual curiosity and willingness for intellectual development among the student teachers.

Online Teaching Practice: When the teaching practice is done in an online environment and through the use of ICT, this type of teaching practice is known as Online Teaching Practice.

Pupil Teachers: The students, who are studying in B.Ed. Course, are called pupil teachers. Pupil teachers are the synonym of student teachers.

Teaching Effectiveness: Teaching Effectiveness can be defined as a manifestation of knowledge of content, skills in lesson presentation, creating a desirable atmosphere for learning and a kind of classroom communications that occur between teachers and students resulting to increase in students' knowledge.

Teaching Practice: In the present study, Teaching Practice refers to the preparation of student teachers for teaching by practical training. The practical training includes teaching methods, teaching strategies, teaching principles and maxims, teaching techniques and related activities learned by the student teachers in daily School life.

Related References

To continue our tradition of advancing information science and technology research, we have compiled a list of recommended IGI Global readings. These references will provide additional information and guidance to further enrich your knowledge and assist you with your own research and future publications.

Acilar, A., & Karamasa, Ç. (2013). Factors affecting e-commerce adoption by small businesses in a developing country: A case study of a small hotel. In S. Chhabra (Ed.), *ICT influences on human development, interaction, and collaboration* (pp. 174–184). Hershey, PA: IGI Global. doi:10.4018/978-1-4666-1957-9.ch010

Ahmed, N. U., Montagno, R., & Sharma, S. (2012). Strategy and structure in a virtual organization. In S. Sharma (Ed.), *E-adoption and technologies for empowering developing countries: Global advances* (pp. 34–45). Hershey, PA: IGI Global. doi:10.4018/978-1-4666-0041-6.ch003

Aikins, S. K., & Chary, M. (2013). Online participation and digital divide: An empirical evaluation of U.S. midwestern municipalities. In *Digital literacy: Concepts, methodologies, tools, and applications* (pp. 63–85). Hershey, PA: IGI Global. doi:10.4018/978-1-4666-1852-7.ch004

Al-Nuaim, H. A. (2013). Developing user profiles for interactive online products in practice. In M. Garcia-Ruiz (Ed.), *Cases on usability engineering: Design and development of digital products* (pp. 57–79). Hershey, PA: IGI Global. doi:10.4018/978-1-4666-4046-7.ch003

Al-Shqairat, Z. I., & Altarawneh, I. I. (2013). The role of partnership in e-government readiness: The knowledge stations (KSs) initiative in Jordan. In A. Mesquita (Ed.), *User perception and influencing factors of technology in everyday life* (pp. 192–210). Hershey, PA: IGI Global. doi:10.4018/978-1-4666-1954-8.ch014

Albers, M. J. (2012). How people approach finding information. In *Human-information interaction and technical communication: Concepts and frameworks* (pp. 398–427). Hershey, PA: IGI Global. doi:10.4018/978-1-4666-0152-9.ch012

Albers, M. J. (2012). How people approach graphical information. In *Human-information interaction and technical communication: Concepts and frameworks* (pp. 262–295). Hershey, PA: IGI Global. doi:10.4018/978-1-4666-0152-9.ch008

Albers, M. J. (2012). How people approach information. In *Human-information interaction and technical communication: Concepts and frameworks* (pp. 114–169). Hershey, PA: IGI Global. doi:10.4018/978-1-4666-0152-9.ch004

Albers, M. J. (2012). How people approach technology-based interactions. In *Human-information interaction and technical communication: Concepts and frameworks* (pp. 170–216). Hershey, PA: IGI Global. doi:10.4018/978-1-4666-0152-9.ch005

Albers, M. J. (2012). How people approach typography. In *Human-information interaction and technical communication: Concepts and frameworks* (pp. 241–261). Hershey, PA: IGI Global. doi:10.4018/978-1-4666-0152-9.ch007

Albers, M. J. (2012). How people interact with information presentation. In *Human-information interaction and technical communication: Concepts and frameworks* (pp. 330–365). Hershey, PA: IGI Global. doi:10.4018/978-1-4666-0152-9.ch010

Albers, M. J. (2012). How people make decisions and take action. In *Human-information interaction and technical communication: Concepts and frameworks* (pp. 428–458). Hershey, PA: IGI Global. doi:10.4018/978-1-4666-0152-9.ch013

Albers, M. J. (2012). How people perform a first glance evaluation. In *Human-information interaction and technical communication: Concepts and frameworks* (pp. 218–240). Hershey, PA: IGI Global. doi:10.4018/978-1-4666-0152-9.ch006

Albers, M. J. (2012). Information in the situation. In *Human-information interaction and technical communication: Concepts and frameworks* (pp. 31–60). Hershey, PA: IGI Global. doi:10.4018/978-1-4666-0152-9.ch002

Albers, M. J. (2012). What people bring with them. In *Human-information interaction and technical communication: Concepts and frameworks* (pp. 61–113). Hershey, PA: IGI Global. doi:10.4018/978-1-4666-0152-9.ch003

Albert, S., & Flournoy, D. (2012). Think global, act local: How ICTs are changing the landscape in community development. In E. Coakes (Ed.), *Technological change and societal growth: Analyzing the future* (pp. 101–116). Hershey, PA: IGI Global. doi:10.4018/978-1-4666-0200-7.ch007

Alexopoulos, G., Koutsouris, A., & Tzouramani, I. (2012). Adoption and use of ICTs among rural youth: Evidence from Greece. In S. Chhabra (Ed.), *ICTs for advancing rural communities and human development: Addressing the digital divide* (pp. 125–142). Hershey, PA: IGI Global. doi:10.4018/978-1-4666-0047-8.ch009

Alfaro, F., Molina, J. P., & Camacho, K. (2012). Public access ICT in Dominican Republic. In R. Gomez (Ed.), *Libraries, telecentres, cybercafes and public access to ICT: International comparisons* (pp. 184–200). Hershey, PA: IGI Global. doi:10.4018/978-1-60960-771-5.ch015

Alhussain, T., & Drew, S. (2012). Employees' perceptions of biometric technology adoption in e-government: An exploratory study in the kingdom of Saudi Arabia. In S. Sharma (Ed.), *E-adoption and technologies for empowering developing countries: Global advances* (pp. 129–142). Hershey, PA: IGI Global. doi:10.4018/978-1-4666-0041-6.ch010

Alkhattabi, M., Neagu, D., & Cullen, A. (2012). User perceptions of information quality in e-learning systems: A gender and cultural perspective. In R. Pande & T. Van der Weide (Eds.), *Globalization, technology diffusion and gender disparity: Social impacts of ICTs* (pp. 138–145). Hershey, PA: IGI Global. doi:10.4018/978-1-4666-0020-1.ch012

Amoretti, F., & Musella, F. (2013). Governing digital divides: Power structures and ICT strategies in a global perspective. In *Digital literacy: Concepts, methodologies, tools, and applications* (pp. 1256–1271). Hershey, PA: IGI Global. doi:10.4018/978-1-4666-1852-7.ch065

Andacht, F. (2013). The tangible lure of the technoself in the age of reality television. In R. Luppicini (Ed.), *Handbook of research on technoself: Identity in a technological society* (pp. 360–381). Hershey, PA: IGI Global. doi:10.4018/978-1-4666-2211-1.ch020

Anderson, C. S., Al-Gahtani, S., & Hubona, G. (2013). The value of TAM antecedents in global IS development and research. In A. Dwivedi & S. Clarke (Eds.), *Innovative strategies and approaches for end-user computing advancements* (pp. 19–39). Hershey, PA: IGI Global. doi:10.4018/978-1-4666-2059-9.ch002

Arifoglu, A., Afacan, G., & Er, E. (2013). Guidelines for successful public internet access points (PIAPs) implementation. In *Digital literacy: Concepts, methodologies, tools, and applications* (pp. 502–521). Hershey, PA: IGI Global. doi:10.4018/978-1-4666-1852-7.ch026

Arning, K., & Ziefle, M. (2012). Ask and you will receive: Training novice adults to use a PDA in an active learning environment. In J. Lumsden (Ed.), *Social and organizational impacts of emerging mobile devices: Evaluating use* (pp. 20–47). Hershey, PA: IGI Global. doi:10.4018/978-1-4666-0194-9.ch002

Arpaci, I., & Gürbüz, T. (2013). Innovation in learning: Innovative tools and techniques for learning. In S. Sharma (Ed.), *Adoption of virtual technologies for business, educational, and governmental advancements* (pp. 117–125). Hershey, PA: IGI Global. doi:10.4018/978-1-4666-2053-7.ch009

Asino, T. I., Wilder, H., & Ferris, S. P. (2012). Innovative use of ICT in Namibia for nationhood: Special emphasis on the Namibian newspaper. In R. Lekoko & L. Semali (Eds.), *Cases on developing countries and ICT integration: Rural community development* (pp. 53–61). Hershey, PA: IGI Global. doi:10.4018/978-1-60960-117-1.ch006

Bainbridge, W. S. (2013). Ancestor veneration avatars. In R. Luppicini (Ed.), *Handbook of research on technoself: Identity in a technological society* (pp. 308–321). Hershey, PA: IGI Global. doi:10.4018/978-1-4666-2211-1.ch017

Baker, P. M., Fairchild, A., & Pater, J. (2012). E-accessibility and municipal wi-fi: Exploring a model for inclusivity and implementation. In S. Chhabra (Ed.), *ICTs for advancing rural communities and human development: Addressing the digital divide* (pp. 109–124). Hershey, PA: IGI Global. doi:10.4018/978-1-4666-0047-8.ch008

Ballesté, F., & Torras, C. (2013). Effects of human-machine integration on the construction of identity. In R. Luppicini (Ed.), *Handbook of research on technoself: Identity in a technological society* (pp. 574–591). Hershey, PA: IGI Global. doi:10.4018/978-1-4666-2211-1.ch030

Barker, R. (2013). Social networking and identity. In R. Luppicini (Ed.), *Handbook of research on technoself: Identity in a technological society* (pp. 474–501). Hershey, PA: IGI Global. doi:10.4018/978-1-4666-2211-1.ch026

Barker, S., & Fiedler, B. (2013). Developers, decision makers, strategists or just end-users? Redefining end-user computing for the 21st century: A case study. In A. Dwivedi & S. Clarke (Eds.), *Innovative strategies and approaches for end-user computing advancements* (pp. 61–76). Hershey, PA: IGI Global. doi:10.4018/978-1-4666-2059-9.ch004

Barlach, A., Hertzum, M., & Simonsen, J. (2013). Pilot implementation driven by effects specifications and formative usability evaluation. In M. Garcia-Ruiz (Ed.), *Cases on usability engineering: Design and development of digital products* (pp. 221–254). Hershey, PA: IGI Global. doi:10.4018/978-1-4666-4046-7.ch010

Barón, L. F., & Valdés, M. (2012). Public access ICT in Colombia. In R. Gomez (Ed.), *Libraries, telecentres, cybercafes and public access to ICT: International comparisons* (pp. 169–183). Hershey, PA: IGI Global. doi:10.4018/978-1-60960-771-5.ch014

Baum, S., & Mahizhnan, A. (2013). Public participation in e-government: Some questions about social inclusion in the Singapore model. In *Digital literacy: Concepts, methodologies, tools, and applications* (pp. 1044–1058). Hershey, PA: IGI Global. doi:10.4018/978-1-4666-1852-7.ch054

Beedle, J., & Wang, S. (2013). Roles of a technology leader. In S. Wang & T. Hartsell (Eds.), *Technology integration and foundations for effective leadership* (pp. 228–241). Hershey, PA: IGI Global. doi:10.4018/978-1-4666-2656-0.ch013

Beklemishev, A. P. (2012). Public access ICT in Kazakhstan. In R. Gomez (Ed.), *Libraries, telecentres, cybercafes and public access to ICT: International comparisons* (pp. 330–343). Hershey, PA: IGI Global. doi:10.4018/978-1-60960-771-5.ch024

Ben, E. R. (2012). Gendering professionalism in the internationalization of information work. In R. Pande & T. Van der Weide (Eds.), *Globalization, technology diffusion and gender disparity: Social impacts of ICTs* (pp. 51–69). Hershey, PA: IGI Global. doi:10.4018/978-1-4666-0020-1.ch005

Bentley, C. M. (2013). Designing and implementing online collaboration tools in West Africa. In N. Azab (Ed.), *Cases on web 2.0 in developing countries: Studies on implementation, application, and use* (pp. 33–60). Hershey, PA: IGI Global. doi:10.4018/978-1-4666-2515-0.ch002

Bentley, Y., & Clarke, S. (2013). Evaluation of information strategy implementation: A critical approach. In A. Dwivedi & S. Clarke (Eds.), *Innovative strategies and approaches for end-user computing advancements* (pp. 1–18). Hershey, PA: IGI Global. doi:10.4018/978-1-4666-2059-9.ch001

Beuschel, W. (2013). The net generation and changes in knowledge acquisition. In T. Takševa (Ed.), *Social software and the evolution of user expertise: Future trends in knowledge creation and dissemination* (pp. 201–226). Hershey, PA: IGI Global. doi:10.4018/978-1-4666-2178-7.ch012

Bhaskar, S. (2012). Moving with time and strategy: India and Bangladesh's development in the era of ICTs. In R. Lekoko & L. Semali (Eds.), *Cases on developing countries and ICT integration: Rural community development* (pp. 146–161). Hershey, PA: IGI Global. doi:10.4018/978-1-60960-117-1.ch014

Bielenia-Grajewska, M. (2013). Actor-network-theory in medical e-communication: The role of websites in creating and maintaining healthcare corporate online identity. In A. Tatnall (Ed.), *Social and professional applications of actor-network theory for technology development* (pp. 156–172). Hershey, PA: IGI Global. doi:10.4018/978-1-4666-2166-4.ch013

Black, D. (2013). The digital soul. In R. Luppicini (Ed.), *Handbook of research on technoself: Identity in a technological society* (pp. 157–174). Hershey, PA: IGI Global. doi:10.4018/978-1-4666-2211-1.ch009

Boelter, J., & Kaschub, C. (2013). Developing the intel® pair & share experience. In M. Garcia-Ruiz (Ed.), *Cases on usability engineering: Design and development of digital products* (pp. 171–194). Hershey, PA: IGI Global. doi:10.4018/978-1-4666-4046-7.ch008

Bonelli, S., & Napoletano, L. (2013). The usability evaluation of a touch screen in the flight deck. In M. Garcia-Ruiz (Ed.), *Cases on usability engineering: Design and development of digital products* (pp. 270–297). Hershey, PA: IGI Global. doi:10.4018/978-1-4666-4046-7.ch012

Bourgault, M., Drouin, N., Sicotte, H., & Daoudi, J. (2012). Moderating effect of team distributedness on organizational dimensions for innovation project success. In A. Mesquita (Ed.), *Human interaction with technology for working, communicating, and learning: Advancements* (pp. 216–235). Hershey, PA: IGI Global. doi:10.4018/978-1-61350-465-9.ch014

Bowe, B. J., Blom, R., & Freedman, E. (2013). Negotiating boundaries between control and dissent: Free speech, business, and repressitarian governments. In J. Lannon & E. Halpin (Eds.), *Human rights and information communication technologies: Trends and consequences of use* (pp. 36–55). Hershey, PA: IGI Global. doi:10.4018/978-1-4666-1918-0.ch003

Bricout, J. C., & Baker, P. M. (2012). Deploying information and communication technologies (ICT) to enhance participation in local governance for citizens with disabilities. In S. Chhabra (Ed.), *ICTs for advancing rural communities and human development: Addressing the digital divide* (pp. 91–108). Hershey, PA: IGI Global. doi:10.4018/978-1-4666-0047-8.ch007

Butcher, K. R., & Sumner, T. (2013). How does prior knowledge impact students' online learning behaviors? In R. Zheng (Ed.), *Evolving psychological and educational perspectives on cyber behavior* (pp. 97–115). Hershey, PA: IGI Global. doi:10.4018/978-1-4666-1858-9.ch007

Bwalya, K. J. (2012). Botswana's novel approaches for knowledge-based economy facilitation: Issues, policies and contextual framework. In S. Chhabra (Ed.), *ICTs for advancing rural communities and human development: Addressing the digital divide* (pp. 45–56). Hershey, PA: IGI Global. doi:10.4018/978-1-4666-0047-8.ch004

Bwalya, K. J. (2012). Towards a knowledge-based economy – The case of Botswana: A discussion article. In E. Coakes (Ed.), *Technological change and societal growth: Analyzing the future* (pp. 117–127). Hershey, PA: IGI Global. doi:10.4018/978-1-4666-0200-7.ch008

Bwalya, K. J., Du Plessis, T., & Rensleigh, C. (2013). A snapshot overview of the digital divide: E-inclusion and e-government in the Zambian context. In *Digital literacy: Concepts, methodologies, tools, and applications* (pp. 1–19). Hershey, PA: IGI Global. doi:10.4018/978-1-4666-1852-7.ch001

Byrd, T. A., & Byrd, L. W. (2012). Contrasting IT capability and organizational types: Implications for firm performance. In A. Dwivedi & S. Clarke (Eds.), *End-user computing, development, and software engineering: New challenges* (pp. 1–24). Hershey, PA: IGI Global. doi:10.4018/978-1-4666-0140-6.ch001

Cabitza, F., & Simone, C. (2012). WOAD: A framework to enable the end-user development of coordination-oriented functionalities. In A. Dwivedi & S. Clarke (Eds.), *End-user computing, development, and software engineering: New challenges* (pp. 127–147). Hershey, PA: IGI Global. doi:10.4018/978-1-4666-0140-6.ch006

Carofiglio, V., & Abbattista, F. (2013). BCI-based user-centered design for emotionally-driven user experience. In M. Garcia-Ruiz (Ed.), *Cases on usability engineering: Design and development of digital products* (pp. 299–320). Hershey, PA: IGI Global. doi:10.4018/978-1-4666-4046-7.ch013

Carpenter, J. (2013). Just doesn't look right: Exploring the impact of humanoid robot integration into explosive ordnance disposal teams. In R. Luppicini (Ed.), *Handbook of research on technoself: Identity in a technological society* (pp. 609–636). Hershey, PA: IGI Global. doi:10.4018/978-1-4666-2211-1.ch032

Cartelli, A. (2013). A framework for digital competence assessment. In *Digital literacy: Concepts, methodologies, tools, and applications* (pp. 196–212). Hershey, PA: IGI Global. doi:10.4018/978-1-4666-1852-7.ch011

Cartelli, A. (2013). Theory and practice in digital competence assessment. In *Digital literacy: Concepts, methodologies, tools, and applications* (pp. 1367–1383). Hershey, PA: IGI Global. doi:10.4018/978-1-4666-1852-7.ch072

Carter, P. D. (2012). The emerging story of the machine. In A. Mesquita (Ed.), *Human interaction with technology for working, communicating, and learning: advancements* (pp. 1–12). Hershey, PA: IGI Global. doi:10.4018/978-1-61350-465-9.ch001

Chadwick, D. D., Fullwood, C., & Wesson, C. J. (2013). Intellectual disability, identity, and the internet. In R. Luppicini (Ed.), *Handbook of research on technoself: Identity in a technological society* (pp. 229–254). Hershey, PA: IGI Global. doi:10.4018/978-1-4666-2211-1.ch013

Chaka, C. (2013). Digitization and consumerization of identity, culture, and power among gen mobinets in South Africa. In R. Luppicini (Ed.), *Handbook of research on technoself: Identity in a technological society* (pp. 399–418). Hershey, PA: IGI Global. doi:10.4018/978-1-4666-2211-1.ch022

Chan, L., & Daim, T. (2012). High technology industrialization and internationalization: Exploring international technology transfer. In A. Cakir & P. Ordóñez de Pablos (Eds.), *Social development and high technology industries: Strategies and applications* (pp. 70–98). Hershey, PA: IGI Global. doi:10.4018/978-1-61350-192-4.ch006

Chand, A. (2013). Reducing digital divide: The case of the 'people first network' (PFNet) in the Solomon Islands. In *Digital literacy: Concepts, methodologies, tools, and applications* (pp. 1571–1605). Hershey, PA: IGI Global. doi:10.4018/978-1-4666-1852-7.ch083

Charnkit, P., & Tatnall, A. (2013). Knowledge conversion processes in Thai public organisations seen as an innovation: The re-analysis of a TAM study using innovation translation. In A. Tatnall (Ed.), *Social and professional applications of actor-network theory for technology development* (pp. 88–102). Hershey, PA: IGI Global. doi:10.4018/978-1-4666-2166-4.ch008

Chary, M. (2013). Social equity, the digital divide and e-governance: An analysis of e-governance initiatives in India. In *Digital literacy: Concepts, methodologies, tools, and applications* (pp. 1321–1334). Hershey, PA: IGI Global. doi:10.4018/978-1-4666-1852-7.ch069

Chary, M., & Aikins, S. K. (2013). Policy as a bridge across the global digital divide. In *Digital literacy: Concepts, methodologies, tools, and applications* (pp. 364–379). Hershey, PA: IGI Global. doi:10.4018/978-1-4666-1852-7.ch019

Chase, N. M., & Clegg, B. (2013). Effects of email utilization on higher education professionals. In A. Mesquita (Ed.), *User perception and influencing factors of technology in everyday life* (pp. 233–247). Hershey, PA: IGI Global. doi:10.4018/978-1-4666-1954-8.ch016

Chen, T., Harper, S., & Yesilada, Y. (2013). How do people use their mobile phones? A field study of small device users. In J. Lumsden (Ed.), *Developments in technologies for human-centric mobile computing and applications* (pp. 38–55). Hershey, PA: IGI Global. doi:10.4018/978-1-4666-2068-1.ch003

Chen, W., & Xie, W. (2012). Cyber behaviors of immigrants. In Z. Yan (Ed.), *Encyclopedia of cyber behavior* (pp. 259–272). Hershey, PA: IGI Global. doi:10.4018/978-1-4666-0315-8.ch022

Chen, Y. (2012). The center for mobile communication studies. In Z. Yan (Ed.), *Encyclopedia of cyber behavior* (pp. 77–87). Hershey, PA: IGI Global. doi:10.4018/978-1-4666-0315-8.ch006

Chen, Y., Lee, B., & Kirk, R. M. (2013). Internet use among older adults: Constraints and opportunities. In R. Zheng, R. Hill, & M. Gardner (Eds.), *Engaging older adults with modern technology: Internet use and information access needs* (pp. 124–141). Hershey, PA: IGI Global. doi:10.4018/978-1-4666-1966-1.ch007

Chilana, P. K., Fishman, E., Geraghty, E. M., Tarczy-Hornoch, P., Wolf, F. M., & Anderson, N. R. (2013). Characterizing data discovery and end-user computing needs in clinical translational science. In A. Dwivedi & S. Clarke (Eds.), *Innovative strategies and approaches for end-user computing advancements* (pp. 301–313). Hershey, PA: IGI Global. doi:10.4018/978-1-4666-2059-9.ch016

Choi, S. M., Chu, S., & Kim, Y. (2012). Culture-laden social engagement: A comparative study of social relationships in social networking sites among American, Chinese and Korean Users. In K. St.Amant & S. Kelsey (Eds.), *Computer-mediated communication across cultures: International interactions in online environments* (pp. 1–16). Hershey, PA: IGI Global. doi:10.4018/978-1-60960-833-0.ch001

Chong, L., Mei, W., & Guang, A. W. (2013). Research and implementation of self-publishing website platforms for universities based on CMS. In T. Gao (Ed.), *Global applications of pervasive and ubiquitous computing* (pp. 166–178). Hershey, PA: IGI Global. doi:10.4018/978-1-4666-2645-4.ch019

Chudoba, K. M., Watson-Manheim, M. B., Crowston, K., & Lee, C. S. (2013). Participation in ICT-enabled meetings. In A. Dwivedi & S. Clarke (Eds.), *Innovative strategies and approaches for end-user computing advancements* (pp. 192–214). Hershey, PA: IGI Global. doi:10.4018/978-1-4666-2059-9.ch011

Clark, E. (2013). Interaction and expertise in an Appalachian music archive. In T. Takševa (Ed.), *Social software and the evolution of user expertise: Future trends in knowledge creation and dissemination* (pp. 311–329). Hershey, PA: IGI Global. doi:10.4018/978-1-4666-2178-7.ch018

Coakes, E., & Coakes, J. (2012). Exploring meaning: The implications of a hyphen for socio-technical theory and practice. In E. Coakes (Ed.), *Technological change and societal growth: Analyzing the future* (pp. 1–35). Hershey, PA: IGI Global. doi:10.4018/978-1-4666-0200-7.ch001

Colomo-Palacios, R., Ruano-Mayoral, M., Soto-Acosta, P., & García-Crespo, Á. (2012). Identifying competences in IT professionals through semantics. In E. Coakes (Ed.), *Technological change and societal growth: Analyzing the future* (pp. 88–99). Hershey, PA: IGI Global. doi:10.4018/978-1-4666-0200-7.ch006

Comunello, F. (2013). From the digital divide to multiple divides: Technology, society, and new media skills. In *Digital literacy: Concepts, methodologies, tools, and applications* (pp. 1622–1639). Hershey, PA: IGI Global. doi:10.4018/978-1-4666-1852-7.ch085

Concha, A. S. (2012). Filipino cyborg sexualities, chatroom masculinities, self-ascribed identities, ephemeral selves. In R. Pande & T. Van der Weide (Eds.), *Globalization, technology diffusion and gender disparity: Social impacts of ICTs* (pp. 211–224). Hershey, PA: IGI Global. doi:10.4018/978-1-4666-0020-1.ch018

Conway, J. (2013). Getting lost in the labyrinth: Information and technology in the marketplace. In B. Medlin (Ed.), *Integrations of technology utilization and social dynamics in organizations* (pp. 226–242). Hershey, PA: IGI Global. doi:10.4018/978-1-4666-1948-7.ch014

Corbett, J., & Mann, R. (2012). Tlowitsis re-imagined: The use of digital media to build nation and overcome disconnection in a displaced aboriginal community. In S. Chhabra (Ed.), *ICTs for advancing rural communities and human development: Addressing the DIGITAL DIVIDE* (pp. 158–179). Hershey, PA: IGI Global. doi:10.4018/978-1-4666-0047-8.ch011

Cordella, A. (2013). Emerging standardization. In A. Tatnall (Ed.), *Social and professional applications of actor-network theory for technology development* (pp. 221–237). Hershey, PA: IGI Global. doi:10.4018/978-1-4666-2166-4.ch017

Corrocher, N., & Raineri, A. (2013). The evolution of the digital divide across developing countries: Theoretical issues and empirical investigation. In *Digital literacy: Concepts, methodologies, tools, and applications* (pp. 1554–1570). Hershey, PA: IGI Global. doi:10.4018/978-1-4666-1852-7.ch082

Coulton, P., & Bamford, W. (2013). Experimenting through mobile 'apps' and 'app stores. In J. Lumsden (Ed.), *Developments in technologies for human-centric mobile computing and applications* (pp. 277–293). Hershey, PA: IGI Global. doi:10.4018/978-1-4666-2068-1.ch017

Cropf, R. A., Benmamoun, M., & Kalliny, M. (2013). The role of web 2.0 in the Arab spring. In N. Azab (Ed.), *Cases on web 2.0 in developing countries: Studies on implementation, application, and use* (pp. 76–108). Hershey, PA: IGI Global. doi:10.4018/978-1-4666-2515-0.ch004

D'Andrea, A., Ferri, F., & Grifoni, P. (2013). Assessing e-health in Africa: Web 2.0 applications. In N. Azab (Ed.), *Cases on web 2.0 in developing countries: Studies on implementation, application, and use* (pp. 442–467). Hershey, PA: IGI Global. doi:10.4018/978-1-4666-2515-0.ch016

DaCosta, B., Kinsell, C., & Nasah, A. (2013). Millennials are digital natives? An investigation into digital propensity and age. In *Digital literacy: Concepts, methodologies, tools, and applications* (pp. 103–119). Hershey, PA: IGI Global. doi:10.4018/978-1-4666-1852-7.ch006

Dahms, M. (2013). Shifting focus from access to impact: Can computers alleviate poverty? In *Digital literacy: Concepts, methodologies, tools, and applications* (pp. 1743–1770). Hershey, PA: IGI Global. doi:10.4018/978-1-4666-1852-7.ch091

Danihelka, J., Hak, R., Kencl, L., & Zara, J. (2013). 3D talking-head interface to voice-interactive services on mobile phones. In J. Lumsden (Ed.), *Developments in technologies for human-centric mobile computing and applications* (pp. 130–144). Hershey, PA: IGI Global. doi:10.4018/978-1-4666-2068-1.ch008

Daugherty, J. L., Mentzer, N. J., Lybrook, D. O., & Little-Wiles, J. (2013). Philosophical perspectives on technology leadership. In S. Wang & T. Hartsell (Eds.), *Technology integration and foundations for effective leadership* (pp. 42–56). Hershey, PA: IGI Global. doi:10.4018/978-1-4666-2656-0.ch003

Dedeaux, T. (2013). An introduction to educational research. In S. Wang & T. Hartsell (Eds.), *Technology integration and foundations for effective leadership* (pp. 319–340). Hershey, PA: IGI Global. doi:10.4018/978-1-4666-2656-0.ch019

Deokar, A. V., Meservy, T. O., Helquist, J. H., & Kruse, J. (2013). Understanding collaboration success in context of cognitive and social presence. In B. Medlin (Ed.), *Integrations of technology utilization and social dynamics in organizations* (pp. 91–107). Hershey, PA: IGI Global. doi:10.4018/978-1-4666-1948-7.ch006

DiMarco, J. (2013). Implementing a website portal using portfoliovillage.com to evaluate professional credentials. In S. Wang & T. Hartsell (Eds.), *Technology integration and foundations for effective leadership* (pp. 308–317). Hershey, PA: IGI Global. doi:10.4018/978-1-4666-2656-0.ch018

Doll, W. J., & Deng, X. (2013). Antecedents of improvisation in IT-enabled engineering work. In A. Dwivedi & S. Clarke (Eds.), *Innovative strategies and approaches for end-user computing advancements* (pp. 242–264). Hershey, PA: IGI Global. doi:10.4018/978-1-4666-2059-9.ch013

Doyle, J., Bertolotto, M., & Wilson, D. (2013). Towards multimodal mobile GIS for the elderly. In *Digital literacy: Concepts, methodologies, tools, and applications* (pp. 590–609). Hershey, PA: IGI Global. doi:10.4018/978-1-4666-1852-7.ch031

Drigas, A., Kouremenos, D., & Vrettaros, J. (2013). Learning applications for disabled people. In *Digital literacy: Concepts, methodologies, tools, and applications* (pp. 1090–1103). Hershey, PA: IGI Global. doi:10.4018/978-1-4666-1852-7.ch057

Dunn, H. S. (2013). Information literacy and the digital divide: Challenging e-exclusion in the global south. In *Digital literacy: Concepts, methodologies, tools, and applications* (pp. 20–38). Hershey, PA: IGI Global. doi:10.4018/978-1-4666-1852-7.ch002

Dunn, R. A. (2013). Identity theories and technology. In R. Luppicini (Ed.), *Handbook of research on technoself: Identity in a technological society* (pp. 26–44). Hershey, PA: IGI Global. doi:10.4018/978-1-4666-2211-1.ch002

Edenius, M., & Rämö, H. (2013). An office on the go: Professional workers, smartphones and the return of place. In A. Mesquita (Ed.), *User perception and influencing factors of technology in everyday life* (pp. 158–177). Hershey, PA: IGI Global. doi:10.4018/978-1-4666-1954-8.ch012

Edwards, R. (2013). Pogo chat. In M. Garcia-Ruiz (Ed.), *Cases on usability engineering: Design and development of digital products* (pp. 378–404). Hershey, PA: IGI Global. doi:10.4018/978-1-4666-4046-7.ch016

Eijkman, H., Herrmann, A., & Savige, K. (2012). E-assessment as a driver for cultural change in network-centric learning. In S. Sharma (Ed.), *E-adoption and technologies for empowering developing countries: Global advances* (pp. 233–244). Hershey, PA: IGI Global. doi:10.4018/978-1-4666-0041-6.ch016

Eisenberg, M., Buechley, L., & Elumeze, N. (2012). Bits and pieces: Potential future scenarios for children's mobile technology. In J. Lumsden (Ed.), *Social and organizational impacts of emerging mobile devices: Evaluating use* (pp. 108–123). Hershey, PA: IGI Global. doi:10.4018/978-1-4666-0194-9.ch006

El Ali, A., Nack, F., & Hardman, L. (2013). Good times?! 3 problems and design considerations for playful HCI. In J. Lumsden (Ed.), *Developments in technologies for human-centric mobile computing and applications* (pp. 204–221). Hershey, PA: IGI Global. doi:10.4018/978-1-4666-2068-1.ch013

Elias, N. (2013). Immigrants' internet use and identity from an intergenerational perspective: Immigrant senior citizens and youngsters from the former Soviet Union in Israel. In R. Luppicini (Ed.), *Handbook of research on technoself: Identity in a technological society* (pp. 293–307). Hershey, PA: IGI Global. doi:10.4018/978-1-4666-2211-1.ch016

Encuentra, E. H., Fernández, M. P., & Gómez-Zúñiga, B. (2013). The internet and older adults: Initial adoption and experience of use. In R. Zheng, R. Hill, & M. Gardner (Eds.), *Engaging older adults with modern technology: Internet use and information access needs* (pp. 212–228). Hershey, PA: IGI Global. doi:10.4018/978-1-4666-1966-1.ch011

Eslambolchilar, P., & Murray-Smith, R. (2012). A model-based approach to analysis and calibration of sensor-based human interaction loops. In J. Lumsden (Ed.), *Social and organizational impacts of emerging mobile devices: Evaluating use* (pp. 48–71). Hershey, PA: IGI Global. doi:10.4018/978-1-4666-0194-9.ch003

Evans, C., & Palacios, L. (2013). Interactive self-assessment questions within a virtual environment. In S. Sharma (Ed.), *Adoption of virtual technologies for business, educational, and governmental advancements* (pp. 218–227). Hershey, PA: IGI Global. doi:10.4018/978-1-4666-2053-7.ch018

Fallery, B., Taddei, R., & Gerbaix, S. (2012). Acceptance and appropriation of videoconferencing for e-training: An empirical investigation. In A. Mesquita (Ed.), *Human interaction with technology for working, communicating, and learning: Advancements* (pp. 104–119). Hershey, PA: IGI Global. doi:10.4018/978-1-61350-465-9.ch007

Farmer, L. (2013). New perspectives of andragogy in relation to the use of technology. In *Digital literacy: Concepts, methodologies, tools, and applications* (pp. 1606–1621). Hershey, PA: IGI Global. doi:10.4018/978-1-4666-1852-7.ch084

Fenley, S. (2013). Navigation and visualisation techniques in elearning and internet research. In *Digital literacy: Concepts, methodologies, tools, and applications* (pp. 636–667). Hershey, PA: IGI Global. doi:10.4018/978-1-4666-1852-7.ch033

Fidler, C. S., Kanaan, R. K., & Rogerson, S. (2013). Barriers to e-government implementation in Jordan: The role of Wasta. In A. Mesquita (Ed.), *User perception and influencing factors of technology in everyday life* (pp. 179–191). Hershey, PA: IGI Global. doi:10.4018/978-1-4666-1954-8.ch013

Fitzgibbons, M. (2013). Teaching political science students to find and evaluate information in the social media flow. In T. Takševa (Ed.), *Social software and the evolution of user expertise: Future trends in knowledge creation and dissemination* (pp. 180–200). Hershey, PA: IGI Global. doi:10.4018/978-1-4666-2178-7.ch011

Flammia, M. (2012). Using the cultural challenges of virtual team projects to prepare students for global citizenship. In K. St.Amant & S. Kelsey (Eds.), *Computer-mediated communication across cultures: International interactions in online environments* (pp. 328–343). Hershey, PA: IGI Global. doi:10.4018/978-1-60960-833-0.ch021

Folorunso, O. O., Vincent, R. O., Ogunde, A. A., & Agboola, B. (2012). Knowledge sharing adoption model based on artificial neural networks. In S. Sharma (Ed.), *E-adoption and technologies for empowering developing countries: Global advances* (pp. 46–58). Hershey, PA: IGI Global. doi:10.4018/978-1-4666-0041-6.ch004

Fors, A. C. (2013). The ontology of the subject in digitalization. In R. Luppicini (Ed.), *Handbook of research on technoself: Identity in a technological society* (pp. 45–63). Hershey, PA: IGI Global. doi:10.4018/978-1-4666-2211-1.ch003

Franckel, S., Bonsignore, E., & Druin, A. (2012). Designing for children's mobile storytelling. In J. Lumsden (Ed.), *Social and organizational impacts of emerging mobile devices: Evaluating use* (pp. 90–107). Hershey, PA: IGI Global. doi:10.4018/978-1-4666-0194-9.ch005

Frobish, T. S. (2013). On pixels, perceptions, and personae: Toward a model of online ethos. In M. Folk & S. Apostel (Eds.), *Online credibility and digital ethos: Evaluating computer-mediated communication* (pp. 1–23). Hershey, PA: IGI Global. doi:10.4018/978-1-4666-2663-8.ch001

Furtmüller, E., Wilderom, C., & van Dick, R. (2012). Sustainable e-recruiting portals: How to motivate applicants to stay connected throughout their careers? In A. Mesquita (Ed.), *Human interaction with technology for working, communicating, and learning: Advancements* (pp. 66–86). Hershey, PA: IGI Global. doi:10.4018/978-1-61350-465-9.ch005

García-Gómez, A. (2013). Technoself-presentation on social networks: A gender-based approach. In R. Luppicini (Ed.), *Handbook of research on technoself: Identity in a technological society* (pp. 382–398). Hershey, PA: IGI Global. doi:10.4018/978-1-4666-2211-1.ch021

Garofalakis, J., & Koskeris, A. (2013). Digital divide and rural communities: Practical solutions and policies. In *Digital literacy: Concepts, methodologies, tools, and applications* (pp. 698–720). Hershey, PA: IGI Global. doi:10.4018/978-1-4666-1852-7.ch036

Gauzente, C. (2012). Does anybody read SMS-advertising? A qualitative and quantitative study of mobile users' attitudes and perceived ad-clutter. In A. Mesquita (Ed.), *Human interaction with technology for working, communicating, and learning: Advancements* (pp. 13–30). Hershey, PA: IGI Global. doi:10.4018/978-1-61350-465-9.ch002

Germaine-McDaniel, N. S. (2012). The emerging hispanic use of online health information in the United States: Cultural convergence or dissociation? In K. St.Amant & S. Kelsey (Eds.), *Computer-mediated communication across cultures: International interactions in online environments* (pp. 251–265). Hershey, PA: IGI Global. doi:10.4018/978-1-60960-833-0.ch017

Goggins, S., Schmidt, M., Guajardo, J., & Moore, J. L. (2013). 3D virtual worlds: Assessing the experience and informing design. In B. Medlin (Ed.), *Integrations of technology utilization and social dynamics in organizations* (pp. 194–213). Hershey, PA: IGI Global. doi:10.4018/978-1-4666-1948-7.ch012

Gomez, R. (2012). Success factors for public access computing: Beyond anecdotes of success. In R. Gomez (Ed.), *Libraries, telecentres, cybercafes and public access to ICT: International comparisons* (pp. 82–94). Hershey, PA: IGI Global. doi:10.4018/978-1-60960-771-5.ch009

Gomez, R. (2013). Success factors in public access computing for development. In S. Chhabra (Ed.), *ICT influences on human development, interaction, and collaboration* (pp. 97–116). Hershey, PA: IGI Global. doi:10.4018/978-1-4666-1957-9.ch005

Gomez, R., & Camacho, K. (2013). Users of ICT at public access centers: Age, education, gender, and income differences in users. In S. Chhabra (Ed.), *ICT influences on human development, interaction, and collaboration* (pp. 1–21). Hershey, PA: IGI Global. doi:10.4018/978-1-4666-1957-9.ch001

Gong, W. (2012). Government monitoring of online media and its influence on netizens' language use in China. In K. St.Amant & S. Kelsey (Eds.), *Computer-mediated communication across cultures: International interactions in online environments* (pp. 155–172). Hershey, PA: IGI Global. doi:10.4018/978-1-60960-833-0.ch011

Göritz, A. S., Singh, R. K., & Voggeser, B. J. (2012). Human behavior on the WWW. In Z. Yan (Ed.), *Encyclopedia of cyber behavior* (pp. 117–131). Hershey, PA: IGI Global. doi:10.4018/978-1-4666-0315-8.ch010

Gozza-Cohen, M., & May, D. (2012). Individuals with disabilities and internet use. In Z. Yan (Ed.), *Encyclopedia of cyber behavior* (pp. 242–258). Hershey, PA: IGI Global. doi:10.4018/978-1-4666-0315-8.ch021

Grant, A. A. (2013). Textperts: Utilizing students' skills in the teaching of writing. In T. Takševa (Ed.), *Social software and the evolution of user expertise: Future trends in knowledge creation and dissemination* (pp. 247–258). Hershey, PA: IGI Global. doi:10.4018/978-1-4666-2178-7.ch014

Greenhow, C. (2013). Online social networking and learning: What are the interesting research questions? In R. Zheng (Ed.), *Evolving psychological and educational perspectives on cyber behavior* (pp. 15–31). Hershey, PA: IGI Global. doi:10.4018/978-1-4666-1858-9.ch002

Gruich, M. R. (2013). Defining professional development for technology. In S. Wang & T. Hartsell (Eds.), *Technology integration and foundations for effective leadership* (pp. 152–170). Hershey, PA: IGI Global. doi:10.4018/978-1-4666-2656-0.ch009

Grundén, K. (2013). Internal digital divide in organizations. In *Digital literacy: Concepts, methodologies, tools, and applications* (pp. 1104–1117). Hershey, PA: IGI Global. doi:10.4018/978-1-4666-1852-7.ch058

Gu, L., Aiken, M., Wang, J., & Wibowo, K. (2013). The influence of information control upon on-line shopping behavior. In A. Mesquita (Ed.), *User perception and influencing factors of technology in everyday life* (pp. 16–27). Hershey, PA: IGI Global. doi:10.4018/978-1-4666-1954-8.ch003

Guechtouli, M. (2012). E-HRM's impact on an environmental scanning process: How can technology support the selection of information? In A. Mesquita (Ed.), *Human interaction with technology for working, communicating, and learning: Advancements* (pp. 120–133). Hershey, PA: IGI Global. doi:10.4018/978-1-61350-465-9.ch008

Guilloux, V., Laval, F., & Kalika, M. (2012). Theorizing HR intranets: Contextual, strategic and configurative explanations. In A. Mesquita (Ed.), *Human interaction with technology for working, communicating, and learning: Advancements* (pp. 87–103). Hershey, PA: IGI Global. doi:10.4018/978-1-61350-465-9.ch006

Guo, R. X. (2013). A case study of social interaction on ANGEL and student authoring skills. In *Digital literacy: Concepts, methodologies, tools, and applications* (pp. 890–908). Hershey, PA: IGI Global. doi:10.4018/978-1-4666-1852-7.ch046

Hall, L. L., & Johnson, R. D. (2013). Preparing IS students for real-world interaction with end users through service learning: A proposed organizational model. In A. Dwivedi & S. Clarke (Eds.), *Innovative strategies and approaches for end-user computing advancements* (pp. 119–133). Hershey, PA: IGI Global. doi:10.4018/978-1-4666-2059-9.ch007

Han, D., & Braun, K. L. (2013). Promoting active ageing through technology training in Korea. In *Digital literacy: Concepts, methodologies, tools, and applications* (pp. 572–589). Hershey, PA: IGI Global. doi:10.4018/978-1-4666-1852-7.ch030

Hanafizadeh, M. R., Hanafizadeh, P., & Saghaei, A. (2013). Perusing e-readiness and digital divide: From a critical view. In *Digital literacy: Concepts, methodologies, tools, and applications* (pp. 313–346). Hershey, PA: IGI Global. doi:10.4018/978-1-4666-1852-7.ch017

Hanappi-Egger, E. (2012). Exclusiveness vs. inclusiveness in software development: The triple-loop-learning approach. In R. Pande & T. Van der Weide (Eds.), *Globalization, technology diffusion and gender disparity: Social impacts of ICTs* (pp. 96–109). Hershey, PA: IGI Global. doi:10.4018/978-1-4666-0020-1.ch008

Harby, F. A., Qahwaji, R., & Kamala, M. (2012). End-users' acceptance of biometrics authentication to secure e-commerce within the context of Saudi culture: Applying the UTAUT model. In R. Pande & T. Van der Weide (Eds.), *Globalization, technology diffusion and gender disparity: Social impacts of ICTs* (pp. 225–246). Hershey, PA: IGI Global. doi:10.4018/978-1-4666-0020-1.ch019

Hartsell, T., & Wang, S. (2013). Introduction to technology integration and leadership. In S. Wang & T. Hartsell (Eds.), *Technology integration and foundations for effective leadership* (pp. 1–17). Hershey, PA: IGI Global. doi:10.4018/978-1-4666-2656-0.ch001

Hawreliak, J. (2013). "To be shot at without result": Gaming and the rhetoric of immortality. In R. Luppicini (Ed.), *Handbook of research on technoself: Identity in a technological society* (pp. 531–553). Hershey, PA: IGI Global. doi:10.4018/978-1-4666-2211-1.ch028

Heilesen, S. B. (2012). Human behaviors with podcasts. In Z. Yan (Ed.), *Encyclopedia of cyber behavior* (pp. 178–188). Hershey, PA: IGI Global. doi:10.4018/978-1-4666-0315-8.ch015

Heng, J., & Banerji, S. (2013). Low usage of intelligent technologies by the aged: New initiatives to bridge the digital divide. In *Digital literacy: Concepts, methodologies, tools, and applications* (pp. 135–153). Hershey, PA: IGI Global. doi:10.4018/978-1-4666-1852-7.ch008

Henze, N., Pielot, M., Poppinga, B., Schinke, T., & Boll, S. (2013). My app is an experiment: Experience from user studies in mobile app stores. In J. Lumsden (Ed.), *Developments in technologies for human-centric mobile computing and applications* (pp. 294–315). Hershey, PA: IGI Global. doi:10.4018/978-1-4666-2068-1.ch018

Hermeking, M. (2012). Culture, online technology and computer-mediated technical documentation: Contributions from the field of intercultural communication. In K. St.Amant & S. Kelsey (Eds.), *Computer-mediated communication across cultures: International interactions in online environments* (pp. 77–90). Hershey, PA: IGI Global. doi:10.4018/978-1-60960-833-0.ch006

Hester, A. J. (2013). Examining the varying influence of social and technological aspects on adoption and usage of knowledge management systems. In B. Medlin (Ed.), *Integrations of technology utilization and social dynamics in organizations* (pp. 142–158). Hershey, PA: IGI Global. doi:10.4018/978-1-4666-1948-7.ch009

Ho, V. (2013). The need for identity construction in computer-mediated professional communication: A Community of practice perspective. In R. Luppicini (Ed.), *Handbook of research on technoself: Identity in a technological society* (pp. 502–530). Hershey, PA: IGI Global. doi:10.4018/978-1-4666-2211-1.ch027

Hoggan, E. (2012). Crossmodal audio and tactile interaction with mobile touchscreens. In J. Lumsden (Ed.), *Social and organizational impacts of emerging mobile devices: Evaluating use* (pp. 249–264). Hershey, PA: IGI Global. doi:10.4018/978-1-4666-0194-9.ch014

Hsu, C. (2012). Exploring the player flow experience in e-game playing. In A. Mesquita (Ed.), *Human interaction with technology for working, communicating, and learning: Advancements* (pp. 48–65). Hershey, PA: IGI Global. doi:10.4018/978-1-61350-465-9.ch004

Hsu, J., Wang, Z., & Hamilton, K. (2013). Developing and managing digital/technology literacy and effective learning skills in adult learners. In *Digital literacy: Concepts, methodologies, tools, and applications* (pp. 394–413). Hershey, PA: IGI Global. doi:10.4018/978-1-4666-1852-7.ch021

Iannella, R. (2012). Towards e-society policy interoperability for social web networks. In S. Sharma (Ed.), *E-adoption and technologies for empowering developing countries: Global advances* (pp. 83–99). Hershey, PA: IGI Global. doi:10.4018/978-1-4666-0041-6.ch007

Iivari, N. (2012). Culturally compatible usability work: An interpretive case study on the relationship between usability work and its cultural context in software product development organizations. In A. Dwivedi & S. Clarke (Eds.), *End-user computing, development, and software engineering: New challenges* (pp. 54–80). Hershey, PA: IGI Global. doi:10.4018/978-1-4666-0140-6.ch003

Ikolo, V. E. (2013). Gender digital divide and national ICT policies in Africa. In *Digital literacy: Concepts, methodologies, tools, and applications* (pp. 812–832). Hershey, PA: IGI Global. doi:10.4018/978-1-4666-1852-7.ch042

Imran, A., & Gregor, S. (2013). A process model for successful e-government adoption in the least developed countries: A case of Bangladesh. In *Digital literacy: Concepts, methodologies, tools, and applications* (pp. 213–241). Hershey, PA: IGI Global. doi:10.4018/978-1-4666-1852-7.ch012

Ionescu, A. (2013). Cyber identity: Our alter-ego? In R. Luppicini (Ed.), *Handbook of research on technoself: Identity in a technological society* (pp. 189–203). Hershey, PA: IGI Global. doi:10.4018/978-1-4666-2211-1.ch011

Ionescu, A. (2013). ICTs and gender-based rights. In J. Lannon & E. Halpin (Eds.), *Human rights and information communication technologies: Trends and consequences of use* (pp. 214–234). Hershey, PA: IGI Global. doi:10.4018/978-1-4666-1918-0.ch013

Irune, A. A. (2013). Evaluating the visual demand of in-vehicle information systems: The development of a new method. In J. Lumsden (Ed.), *Developments in technologies for human-centric mobile computing and applications* (pp. 1–21). Hershey, PA: IGI Global. doi:10.4018/978-1-4666-2068-1.ch001

Islam, A. N. (2013). Understanding the continued usage intention of educators toward an e-learning system. In S. Sharma (Ed.), *Adoption of virtual technologies for business, educational, and governmental advancements* (pp. 180–195). Hershey, PA: IGI Global. doi:10.4018/978-1-4666-2053-7.ch015

Islam, M. N., & Tétard, F. (2013). Integrating semiotics perception in usability testing to improve usability evaluation. In M. Garcia-Ruiz (Ed.), *Cases on usability engineering: Design and development of digital products* (pp. 145–169). Hershey, PA: IGI Global. doi:10.4018/978-1-4666-4046-7.ch007

Ivanova, M., & Popova, A. (2013). Formal and informal learning flows cohesion in web 2.0 environment. In J. Wang (Ed.), *Information systems and modern society: Social change and global development* (pp. 1–16). Hershey, PA: IGI Global. doi:10.4018/978-1-4666-2922-6.ch001

Iyamu, T. (2013). Institutionalisation of the enterprise architecture: The actor-network perspective. In A. Tatnall (Ed.), *Social and professional applications of actor-network theory for technology development* (pp. 144–155). Hershey, PA: IGI Global. doi:10.4018/978-1-4666-2166-4.ch012

Jahnke, I. (2012). A way out of the information jungle: A longitudinal study about a socio-technical community and informal learning in higher education. In E. Coakes (Ed.), *Technological change and societal growth: Analyzing the future* (pp. 180–201). Hershey, PA: IGI Global. doi:10.4018/978-1-4666-0200-7.ch012

James, T., & Louw, M. (2012). Public access ICT in Namibia. In R. Gomez (Ed.), *Libraries, telecentres, cybercafes and public access to ICT: International comparisons* (pp. 452–465). Hershey, PA: IGI Global. doi:10.4018/978-1-60960-771-5.ch031

Janneck, M., & Staar, H. (2013). Playing virtual power games: Micro-political processes in inter-organizational networks. In B. Medlin (Ed.), *Integrations of technology utilization and social dynamics in organizations* (pp. 171–192). Hershey, PA: IGI Global. doi:10.4018/978-1-4666-1948-7.ch011

Jensen, K. L. (2013). Remote and autonomous studies of mobile and ubiquitous applications in real contexts. In J. Lumsden (Ed.), *Developments in technologies for human-centric mobile computing and applications* (pp. 79–98). Hershey, PA: IGI Global. doi:10.4018/978-1-4666-2068-1.ch005

Jin, L. (2013). A new trend in education: Technoself enhanced social learning. In R. Luppicini (Ed.), *Handbook of research on technoself: Identity in a technological society* (pp. 456–473). Hershey, PA: IGI Global. doi:10.4018/978-1-4666-2211-1.ch025

Johnson, K. T., & Smith-Jackson, T. L. (2013). A human factors view of the digital divide. In *Digital literacy: Concepts, methodologies, tools, and applications* (pp. 1510–1532). Hershey, PA: IGI Global. doi:10.4018/978-1-4666-1852-7.ch080

Johnson, M. (2012). Designing visionary leadership teams. In E. Coakes (Ed.), *Technological change and societal growth: Analyzing the future* (pp. 36–55). Hershey, PA: IGI Global. doi:10.4018/978-1-4666-0200-7.ch002

Johnson, N. (2013). Online credibility and information labor: Infrastructure reverberating through ethos. In M. Folk & S. Apostel (Eds.), *Online credibility and digital ethos: Evaluating computer-mediated communication* (pp. 37–55). Hershey, PA: IGI Global. doi:10.4018/978-1-4666-2663-8.ch003

Johnson, R. D. (2013). Gender differences in e-learning: Communication, social presence, and learning outcomes. In A. Dwivedi & S. Clarke (Eds.), *Innovative strategies and approaches for end-user computing advancements* (pp. 175–191). Hershey, PA: IGI Global. doi:10.4018/978-1-4666-2059-9.ch010

Johnston, A., & Warkentin, M. (2012). The influence of perceived source credibility on end user attitudes and intentions to comply with recommended IT actions. In A. Dwivedi & S. Clarke (Eds.), *End-user computing, development, and software engineering: New challenges* (pp. 312–334). Hershey, PA: IGI Global. doi:10.4018/978-1-4666-0140-6.ch015

Joo, K. P. (2012). ICT-supported education for sustainable development of South Korean rural communities. In R. Lekoko & L. Semali (Eds.), *Cases on developing countries and ICT integration: Rural community development* (pp. 174–191). Hershey, PA: IGI Global. doi:10.4018/978-1-60960-117-1.ch016

Jumisko-Pyykkö, S., & Vainio, T. (2012). Framing the context of use for mobile HCI. In J. Lumsden (Ed.), *Social and organizational impacts of emerging mobile devices: Evaluating use* (pp. 217–248). Hershey, PA: IGI Global. doi:10.4018/978-1-4666-0194-9.ch013

Jyothi, P. (2012). Challenges faced by women: BPO sector. In R. Pande & T. Van der Weide (Eds.), *Globalization, technology diffusion and gender disparity: Social impacts of ICTs* (pp. 147–155). Hershey, PA: IGI Global. doi:10.4018/978-1-4666-0020-1.ch013

Kafai, Y. B., Fields, D., & Burke, W. Q. (2012). Entering the clubhouse: Case studies of young programmers joining the online scratch communities. In A. Dwivedi & S. Clarke (Eds.), *End-user computing, development, and software engineering: New challenges* (pp. 279–294). Hershey, PA: IGI Global. doi:10.4018/978-1-4666-0140-6.ch013

Kaiser, C., & Butcher, G. (2013). Exploring evaluation techniques for children's websites. In M. Garcia-Ruiz (Ed.), *Cases on usability engineering: Design and development of digital products* (pp. 1–25). Hershey, PA: IGI Global. doi:10.4018/978-1-4666-4046-7.ch001

Kamal, S., & Chu, S. (2012). Cultural differences in social media usage and beliefs and attitudes towards advertising on social media: Findings from Dubai, United Arab Emirates. In K. St.Amant & S. Kelsey (Eds.), *Computer-mediated communication across cultures: international interactions in online environments* (pp. 123–141). Hershey, PA: IGI Global. doi:10.4018/978-1-60960-833-0.ch009

Karasavvidis, I. (2013). Rethinking expertise in the web 2.0 era: Lessons learned from project durian. In T. Takševa (Ed.), *Social software and the evolution of user expertise: Future trends in knowledge creation and dissemination* (pp. 330–353). Hershey, PA: IGI Global. doi:10.4018/978-1-4666-2178-7.ch019

Kauppinen, J., Kivijärvi, H., & Talvinen, J. (2013). Committing to organizational change in IT industry. In B. Medlin (Ed.), *Integrations of technology utilization and social dynamics in organizations* (pp. 74–90). Hershey, PA: IGI Global. doi:10.4018/978-1-4666-1948-7.ch005

Kawlra, A. (2013). From rural outsourcing to rural opportunities: Developing an ICT mediated distributed production enterprise in Tamil Nadu, India. In S. Chhabra (Ed.), *ICT influences on human development, interaction, and collaboration* (pp. 158–173). Hershey, PA: IGI Global. doi:10.4018/978-1-4666-1957-9.ch009

Kaynak, M. S. (2013). The potentials and pitfalls of the information society project in Turkey: A critical assessment of policy paradigms regarding e-transformation and digital divide. In *Digital literacy: Concepts, methodologies, tools, and applications* (pp. 833–851). Hershey, PA: IGI Global. doi:10.4018/978-1-4666-1852-7.ch043

Kelly, D. (2013). An analysis of process characteristics for developing scientific software. In A. Dwivedi & S. Clarke (Eds.), *Innovative strategies and approaches for end-user computing advancements* (pp. 347–363). Hershey, PA: IGI Global. doi:10.4018/978-1-4666-2059-9.ch019

Kettani, D., Moulin, B., & El Mahdi, A. (2013). Toward a roadmap to e-government for a better governance toward a roadmap to e-government for a better governance. In *Digital literacy: Concepts, methodologies, tools, and applications* (pp. 1771–1799). Hershey, PA: IGI Global. doi:10.4018/978-1-4666-1852-7.ch092

Khan, G. F., & Moon, J. (2013). E-government issues in developing countries: An analysis from a digital divide, e-skills, and civil conflict theory approach. In *Digital literacy: Concepts, methodologies, tools, and applications* (pp. 1272–1288). Hershey, PA: IGI Global. doi:10.4018/978-1-4666-1852-7.ch066

Klaus, T. (2013). Understanding user dissatisfaction: exploring the role of fairness in IT-enabled change. In A. Dwivedi & S. Clarke (Eds.), *Innovative strategies and approaches for end-user computing advancements* (pp. 215–241). Hershey, PA: IGI Global. doi:10.4018/978-1-4666-2059-9.ch012

Knight, M. L., Knight, R. A., Goben, A., & Dobbs, A. W. (2013). Theory and application: Using social networking to build online credibility. In M. Folk & S. Apostel (Eds.), *Online credibility and digital ethos: Evaluating computer-mediated communication* (pp. 285–301). Hershey, PA: IGI Global. doi:10.4018/978-1-4666-2663-8.ch016

Knoche, H., Rao, P. S., & Huang, J. (2013). Human-centered design for development. In J. Lumsden (Ed.), *Developments in technologies for human-centric mobile computing and applications* (pp. 155–167). Hershey, PA: IGI Global. doi:10.4018/978-1-4666-2068-1.ch010

Korpelainen, E., Vartiainen, M., & Kira, M. (2012). Self-determined adoption of an ICT system in a work organization. In A. Dwivedi & S. Clarke (Eds.), *End-user computing, development, and software engineering: New challenges* (pp. 148–167). Hershey, PA: IGI Global. doi:10.4018/978-1-4666-0140-6.ch007

Korres, G. M., & Kokkinou, A. (2012). Reviewing the European innovation activities and industrial competitiveness. In A. Cakir & P. Ordóñez de Pablos (Eds.), *Social development and high technology industries: Strategies and applications* (pp. 176–186). Hershey, PA: IGI Global. doi:10.4018/978-1-61350-192-4.ch012

Kumar, M., & Sareen, M. (2013). Impact of technology-related environment issues on trust in B2B e-commerce. In S. Chhabra (Ed.), *ICT influences on human development, interaction, and collaboration* (pp. 22–42). Hershey, PA: IGI Global. doi:10.4018/978-1-4666-1957-9.ch002

Kumar, M., Sareen, M., & Chhabra, S. (2013). Technology related trust issues in SME B2B E-Commerce. In S. Chhabra (Ed.), *ICT influences on human development, interaction, and collaboration* (pp. 243–259). Hershey, PA: IGI Global. doi:10.4018/978-1-4666-1957-9.ch015

Kumari, B. R. (2012). Gender, culture, and ICT use. In R. Pande & T. Van der Weide (Eds.), *Globalization, technology diffusion and gender disparity: Social impacts of ICTs* (pp. 36–50). Hershey, PA: IGI Global. doi:10.4018/978-1-4666-0020-1.ch004

Kunal,. (2013). Innovation in financing mechanism of information and communication technology (ICT). In *Digital literacy: Concepts, methodologies, tools, and applications* (pp. 1152-1191). Hershey, PA: IGI Global. doi:10.4018/978-1-4666-1852-7.ch061

Kvasny, L., & Hales, K. D. (2013). The evolving discourse of the digital divide: The internet, black identity, and the evolving discourse of the digital divide. In *Digital literacy: Concepts, methodologies, tools, and applications* (pp. 1350–1366). Hershey, PA: IGI Global. doi:10.4018/978-1-4666-1852-7.ch071

L'Abate, L. (2013). Of paradigms, theories, and models: A conceptual hierarchical structure for communication science and technoself. In R. Luppicini (Ed.), *Handbook of research on technoself: Identity in a technological society* (pp. 84–104). Hershey, PA: IGI Global. doi:10.4018/978-1-4666-2211-1.ch005

Langran, E. (2013). Caste, class, and IT in India. In *Digital literacy: Concepts, methodologies, tools, and applications* (pp. 976–994). Hershey, PA: IGI Global. doi:10.4018/978-1-4666-1852-7.ch050

Larsen, J. E., & Stopczynski, A. (2013). A festival-wide social network using 2D barcodes, mobile phones and situated displays. In J. Lumsden (Ed.), *Developments in technologies for human-centric mobile computing and applications* (pp. 168–185). Hershey, PA: IGI Global. doi:10.4018/978-1-4666-2068-1.ch011

Lawrence, H. R. (2013). Women's roles: Do they exist in a technological workforce? In S. Wang & T. Hartsell (Eds.), *Technology integration and foundations for effective leadership* (pp. 57–69). Hershey, PA: IGI Global. doi:10.4018/978-1-4666-2656-0.ch004

Laxman, K., & Chin, Y. K. (2012). Emergent technologies and social connectedness in learning. In A. Cakir & P. Ordóñez de Pablos (Eds.), *Social development and high technology industries: Strategies and applications* (pp. 25–37). Hershey, PA: IGI Global. doi:10.4018/978-1-61350-192-4.ch003

Lee, J., & Rethemeyer, R. K. (2012). Virtual interactions via smartphones. In Z. Yan (Ed.), *Encyclopedia of cyber behavior* (pp. 189–198). Hershey, PA: IGI Global. doi:10.4018/978-1-4666-0315-8.ch016

Lee, T., Shen, P., & Tsai, C. (2012). Enhance students' computing skills via web-mediated self-regulated learning with feedback in blended environment. In A. Mesquita (Ed.), *Human interaction with technology for working, communicating, and learning: Advancements* (pp. 149–166). Hershey, PA: IGI Global. doi:10.4018/978-1-61350-465-9.ch010

Lekoko, R., Modise-Jankie, J., & Busang, C. (2012). Libraries as portal for knowledge driven rural community development cases from Botswana. In R. Lekoko & L. Semali (Eds.), *Cases on developing countries and ICT integration: Rural community development* (pp. 34–41). Hershey, PA: IGI Global. doi:10.4018/978-1-60960-117-1.ch004

Leichsenring, C., Tünnermann, R., & Hermann, T. (2013). feelabuzz: Direct tactile communication with mobile phones. In J. Lumsden (Ed.), Developments in technologies for human-centric mobile computing and applications (pp. 145-154). Hershey, PA: IGI Global. doi:10.4018/978-1-4666-2068-1.ch009

Leung, L., & Zheng, C. P. (2012). The net generation. In Z. Yan (Ed.), *Encyclopedia of Cyber behavior* (pp. 200–211). Hershey, PA: IGI Global. doi:10.4018/978-1-4666-0315-8.ch017

Lewandowski, D. (2013). Credibility in web search engines. In M. Folk & S. Apostel (Eds.), *Online credibility and digital ethos: Evaluating computer-mediated communication* (pp. 131–146). Hershey, PA: IGI Global. doi:10.4018/978-1-4666-2663-8.ch008

Li, G., Zhang, X., Wang, Z., & Gao, T. (2013). Coordination performance evaluation of supply logistics in JIT environment. In T. Gao (Ed.), *Global applications of pervasive and ubiquitous computing* (pp. 17–30). Hershey, PA: IGI Global. doi:10.4018/978-1-4666-2645-4.ch003

Li, X., & Leong, A. M. (2012). Model analysis and development strategy on building an industrial research and development (R&D) centre: Shanghai's practice and inspiration. In A. Cakir & P. Ordóñez de Pablos (Eds.), *Social development and high technology industries: Strategies and applications* (pp. 1–14). Hershey, PA: IGI Global. doi:10.4018/978-1-61350-192-4.ch001

Li, Y., & Chang, K. (2013). Exploring the dimensions and effects of computer software similarities in computer skills transfer. In A. Dwivedi & S. Clarke (Eds.), *Innovative strategies and approaches for end-user computing advancements* (pp. 99–118). Hershey, PA: IGI Global. doi:10.4018/978-1-4666-2059-9.ch006

Light, A., Kleine, D., Holloway, R., & Vivent, M. (2012). Performing charlotte: A technique to bridge cultures in participatory design. In E. Coakes (Ed.), *Technological change and societal growth: Analyzing the future* (pp. 219–236). Hershey, PA: IGI Global. doi:10.4018/978-1-4666-0200-7.ch014

Linares, K., Subrahmanyam, K., Cheng, R., & Guan, S. A. (2013). A second life within second life: Are virtual world users creating new selves and new lives? In R. Zheng (Ed.), *Evolving psychological and educational perspectives on cyber behavior* (pp. 205–228). Hershey, PA: IGI Global. doi:10.4018/978-1-4666-1858-9.ch013

Lind, L., Berglund, A., Berglund, E., Bång, M., & Hägglund, S. (2013). Effortless data capture for ambient e-services with digital pen and paper technology. In *Digital literacy: Concepts, methodologies, tools, and applications* (pp. 522–540). Hershey, PA: IGI Global. doi:10.4018/978-1-4666-1852-7.ch027

Loewen, S., & Wang, J. (2012). Human behavior in chatrooms. In Z. Yan (Ed.), *Encyclopedia of cyber behavior* (pp. 148–156). Hershey, PA: IGI Global. doi:10.4018/978-1-4666-0315-8.ch012

Low, R., Jin, P., & Sweller, J. (2013). Instructional design in digital environments and availability of mental resources. In R. Zheng, R. Hill, & M. Gardner (Eds.), *Engaging older adults with modern technology: Internet use and information access needs* (pp. 81–104). Hershey, PA: IGI Global. doi:10.4018/978-1-4666-1966-1.ch005

Lukaitis, S. (2013). Applying hermeneutic phenomenology to understand innovation adoption. In A. Tatnall (Ed.), *Social and professional applications of actor-network theory for technology development* (pp. 103–116). Hershey, PA: IGI Global. doi:10.4018/978-1-4666-2166-4.ch009

Luppicini, R. (2013). The emerging field of technoself studies (TSS). In R. Luppicini (Ed.), *Handbook of research on technoself: Identity in a technological society* (pp. 1–25). Hershey, PA: IGI Global. doi:10.4018/978-1-4666-2211-1.ch001

Mabry, S. (2012). Driving IT architecture innovation: The roles of competing organizational cultures and collaborating upper echelons. In S. Sharma (Ed.), *E-adoption and technologies for empowering developing countries: Global advances* (pp. 15–33). Hershey, PA: IGI Global. doi:10.4018/978-1-4666-0041-6.ch002

Mahatanankoon, P. (2012). The impact of personal electronic communications on work-life balance and cognitive absorption. In S. Chhabra (Ed.), *ICTs for advancing rural communities and human development: Addressing the digital divide* (pp. 1–14). Hershey, PA: IGI Global. doi:10.4018/978-1-4666-0047-8.ch001

Maidel, V., & Epstein, D. (2013). The query is just the beginning: Exploring search-related decision-making of young adults. In M. Folk & S. Apostel (Eds.), *Online credibility and digital ethos: Evaluating computer-mediated communication* (pp. 95–113). Hershey, PA: IGI Global. doi:10.4018/978-1-4666-2663-8.ch006

Malleus, R. (2013). Whose news can you trust? A framework for evaluating the credibility of online news sources for diaspora populations. In M. Folk & S. Apostel (Eds.), *Online credibility and digital ethos: Evaluating computer-mediated communication* (pp. 186–214). Hershey, PA: IGI Global. doi:10.4018/978-1-4666-2663-8.ch011

Mano, R. S., & Mesch, G. S. (2012). E-mail and work performance. In Z. Yan (Ed.), *Encyclopedia of cyber behavior* (pp. 106–116). Hershey, PA: IGI Global. doi:10.4018/978-1-4666-0315-8.ch009

Marmura, S. (2013). The mediation of identity: Key issues in historic perspective. In R. Luppicini (Ed.), *Handbook of research on technoself: Identity in a technological society* (pp. 137–156). Hershey, PA: IGI Global. doi:10.4018/978-1-4666-2211-1.ch008

Martin, I., Kear, K., Simpkins, N., & Busvine, J. (2013). Social negotiations in web usability engineering. In M. Garcia-Ruiz (Ed.), *Cases on usability engineering: Design and development of digital products* (pp. 26–56). Hershey, PA: IGI Global. doi:10.4018/978-1-4666-4046-7.ch002

Martin, J., & McKay, E. (2013). Mental health, post-secondary education, and information communications technology. In J. Lannon & E. Halpin (Eds.), *Human rights and information communication technologies: Trends and consequences of use* (pp. 196–213). Hershey, PA: IGI Global. doi:10.4018/978-1-4666-1918-0.ch012

Martin, J. N., & Cheong, P. H. (2012). Cultural considerations of online pedagogy. In K. St.Amant & S. Kelsey (Eds.), *Computer-mediated communication across cultures: International interactions in online environments* (pp. 283–311). Hershey, PA: IGI Global. doi:10.4018/978-1-60960-833-0.ch019

Mattingly, D. J. (2012). Indian women working in call centers: Sites of resistance? In R. Pande & T. Van der Weide (Eds.), *Globalization, technology diffusion and gender disparity: Social impacts of ICTs* (pp. 156–168). Hershey, PA: IGI Global. doi:10.4018/978-1-4666-0020-1.ch014

McColl, D., & Nejat, G. (2013). A human affect recognition system for socially interactive robots. In R. Luppicini (Ed.), *Handbook of research on technoself: Identity in a technological society* (pp. 554–573). Hershey, PA: IGI Global. doi:10.4018/978-1-4666-2211-1.ch029

McDonald, A., & Helmer, S. (2013). A comparative case study of Indonesian and UK organisational culture differences in IS project management. In A. Mesquita (Ed.), *User perception and influencing factors of technology in everyday life* (pp. 46–55). Hershey, PA: IGI Global. doi:10.4018/978-1-4666-1954-8.ch005

McGrady, R. (2013). Ethos [edit]: Procedural rhetoric and the Wikipedia project. In M. Folk, & S. Apostel (Eds.) Online credibility and digital ethos: Evaluating computer-mediated communication (pp. 114-130). Hershey, PA: IGI Global. doi:10.4018/978-1-4666-2663-8.ch007

McLay, A. (2012). Realising virtual reality: A reflection on the continuing evolution of new media. In E. Coakes (Ed.), *Technological change and societal growth: Analyzing the future* (pp. 144–161). Hershey, PA: IGI Global. doi:10.4018/978-1-4666-0200-7.ch010

McMillan, D., Morrison, A., & Chalmers, M. (2013). A comparison of distribution channels for large-scale deployments of iOS applications. In J. Lumsden (Ed.), *Developments in technologies for human-centric mobile computing and applications* (pp. 222–239). Hershey, PA: IGI Global. doi:10.4018/978-1-4666-2068-1.ch014

Melson, G. F. (2013). Building a technoself: Children's ideas about and behavior toward robotic pets. In R. Luppicini (Ed.), *Handbook of research on technoself: Identity in a technological society* (pp. 592–608). Hershey, PA: IGI Global. doi:10.4018/978-1-4666-2211-1.ch031

Mendoza-González, R., Rodríguez, F. Á., & Arteaga, J. M. (2013). A usability study of mobile text based social applications: Towards a reliable strategy for design evaluation. In M. Garcia-Ruiz (Ed.), *Cases on usability engineering: Design and development of digital products* (pp. 195–219). Hershey, PA: IGI Global. doi:10.4018/978-1-4666-4046-7.ch009

Michikyan, M., & Subrahmanyam, K. (2012). Social networking sites: Implications for youth. In Z. Yan (Ed.), *Encyclopedia of cyber behavior* (pp. 132–147). Hershey, PA: IGI Global. doi:10.4018/978-1-4666-0315-8.ch011

Mitra, A. (2013). Collective narrative expertise and the narbs of social media. In T. Takševa (Ed.), *Social software and the evolution of user expertise: Future trends in knowledge creation and dissemination* (pp. 1–12). Hershey, PA: IGI Global. doi:10.4018/978-1-4666-2178-7.ch001

Modise, O. M., Lekoko, R., & Thobega, J. M. (2012). Socio-economic empowerment through technologies: The case of tapestry at Lentswe La Oodi Weavers in Botswana. In R. Lekoko & L. Semali (Eds.), *Cases on developing countries and ICT integration: Rural community development* (pp. 75–82). Hershey, PA: IGI Global. doi:10.4018/978-1-60960-117-1.ch008

Moloi, T. (2012). Linking mathematical literacy to ICT: A good mix for community development in South Africa. In R. Lekoko & L. Semali (Eds.), *Cases on developing countries and ICT integration: Rural community development* (pp. 202–210). Hershey, PA: IGI Global. doi:10.4018/978-1-60960-117-1.ch018

Moore, J. E., & Love, M. S. (2013). An examination of prestigious stigma: The case of the technology geek. In B. Medlin (Ed.), *Integrations of technology utilization and social dynamics in organizations* (pp. 48–73). Hershey, PA: IGI Global. doi:10.4018/978-1-4666-1948-7.ch004

Mørch, A. I., & Andersen, R. (2012). Mutual development: The software engineering context of end-user development. In A. Dwivedi & S. Clarke (Eds.), *End-user computing, development, and software engineering: New challenges* (pp. 103–125). Hershey, PA: IGI Global. doi:10.4018/978-1-4666-0140-6.ch005

Moreau, R. A., & Hershorn, H. R. (2012). Children's internet safety websites. In Z. Yan (Ed.), *Encyclopedia of cyber behavior* (pp. 96–104). Hershey, PA: IGI Global. doi:10.4018/978-1-4666-0315-8.ch008

Moser, S., Bruppacher, S. E., & de Simoni, F. (2013). Public representation of ubiquitous ICT applications in the outpatient health sector. In A. Mesquita (Ed.), *User perception and influencing factors of technology in everyday life* (pp. 212–232). Hershey, PA: IGI Global. doi:10.4018/978-1-4666-1954-8.ch015

Mosindi, O., & Sice, P. (2013). An exploratory theoretical framework for understanding information behaviour. In A. Mesquita (Ed.), *User perception and influencing factors of technology in everyday life* (pp. 1–8). Hershey, PA: IGI Global. doi:10.4018/978-1-4666-1954-8.ch001

Mousten, B., Humbley, J., Maylath, B., & Vandepitte, S. (2012). Communicating pragmatics about content and culture in virtually mediated educational environments. In K. St.Amant & S. Kelsey (Eds.), *Computer-mediated communication across cultures: International interactions in online environments* (pp. 312–327). Hershey, PA: IGI Global. doi:10.4018/978-1-60960-833-0.ch020

Muir, S. A. (2013). The gloss and the reality of teaching digital natives: Taking the long view. In *Digital literacy: Concepts, methodologies, tools, and applications* (pp. 1697–1719). Hershey, PA: IGI Global. doi:10.4018/978-1-4666-1852-7.ch089

Mulcahy, D. (2013). Performativity in practice: An actor-network account of professional teaching standards. In A. Tatnall (Ed.), *Social and professional applications of actor-network theory for technology development* (pp. 1–16). Hershey, PA: IGI Global. doi:10.4018/978-1-4666-2166-4.ch001

Murphy, J., Lee, R., & Swinger, E. (2013). Student perceptions and adoption of university smart card systems. In A. Mesquita (Ed.), *User perception and influencing factors of technology in everyday life* (pp. 142–157). Hershey, PA: IGI Global. doi:10.4018/978-1-4666-1954-8.ch011

Mushi, R. T., & Chilimo, W. (2013). Contribution of information and communication technologies to malaria control in Tanzania. In S. Chhabra (Ed.), *ICT influences on human development, interaction, and collaboration* (pp. 132–141). Hershey, PA: IGI Global. doi:10.4018/978-1-4666-1957-9.ch007

Mutaza, S., & Sami, L. K. (2012). Gender aspects in the use of ICT in information centres. In R. Pande & T. Van der Weide (Eds.), *Globalization, technology diffusion and gender disparity: Social impacts of ICTs* (pp. 129–137). Hershey, PA: IGI Global. doi:10.4018/978-1-4666-0020-1.ch011

Mutula, S. M. (2012). E-government's role in poverty alleviation: Case study of South Africa. In R. Lekoko & L. Semali (Eds.), *Cases on developing countries and ICT integration: Rural community development* (pp. 104–122). Hershey, PA: IGI Global. doi:10.4018/978-1-60960-117-1.ch011

Myers, B. A., Jeong, S. Y., Xie, Y., Beaton, J., Stylos, J., & Ehret, R. ... Busse, D. K. (2012). Studying the documentation of an API for enterprise service-oriented architecture. In A. Dwivedi, & S. Clarke (Eds.) End-user computing, development, and software engineering: New challenges (pp. 81-102). Hershey, PA: IGI Global. doi:10.4018/978-1-4666-0140-6.ch004

Nepali, R. K., & Bista, B. (2012). Public access ICT in Nepal. In R. Gomez (Ed.), *Libraries, telecentres, cybercafes and public access to ICT: International comparisons* (pp. 267–282). Hershey, PA: IGI Global. doi:10.4018/978-1-60960-771-5.ch020

Nezlek, G., & DeHondt, G. (2013). Gender wage differentials in information systems: 1991 – 2008 a quantitative analysis. In B. Medlin (Ed.), *Integrations of technology utilization and social dynamics in organizations* (pp. 31–47). Hershey, PA: IGI Global. doi:10.4018/978-1-4666-1948-7.ch003

Ng, W. (2013). Empowering students to be scientifically literate through digital literacy. In *Digital literacy: Concepts, methodologies, tools, and applications* (pp. 1219–1239). Hershey, PA: IGI Global. doi:10.4018/978-1-4666-1852-7.ch063

Nhlekisana, R. O. (2012). Music as a catalyst for improved livelihood: The case of culture spears, a traditional music dance group in Botswana. In R. Lekoko & L. Semali (Eds.), *Cases on developing countries and ICT integration: Rural community development* (pp. 83–90). Hershey, PA: IGI Global. doi:10.4018/978-1-60960-117-1.ch009

Nicholls, M. G., & Cargill, B. J. (2012). Achieving best practice manufacturing involving tacit knowledge through the cautious use of mixed-mode modelling. In E. Coakes (Ed.), *Technological change and societal growth: Analyzing the future* (pp. 237–256). Hershey, PA: IGI Global. doi:10.4018/978-1-4666-0200-7.ch015

Nordin, K., & Berglund, U. (2012). Children's maps in GIS: A tool for communicating outdoor experiences in urban planning. In S. Chhabra (Ed.), *ICTs for advancing rural communities and human development: Addressing the digital divide* (pp. 57–72). Hershey, PA: IGI Global. doi:10.4018/978-1-4666-0047-8.ch005

Nørgaard, M., & Hornbæk, K. (2012). Working together to improve usability: Exploring challenges and successful practices. In A. Mesquita (Ed.), *Human interaction with technology for working, communicating, and learning: Advancements* (pp. 167–187). Hershey, PA: IGI Global. doi:10.4018/978-1-61350-465-9.ch011

O'Hanlon, S., Bourke, A., & Power, V. (2013). E-health for older adults. In R. Zheng, R. Hill, & M. Gardner (Eds.), *Engaging older adults with modern technology: Internet use and information access needs* (pp. 229–248). Hershey, PA: IGI Global. doi:10.4018/978-1-4666-1966-1.ch012

Oiry, E., Ologeanu-Taddeï, R., & Bondarouk, T. (2012). The role of the organizational structure in the IT appropriation: Explorative case studies into the interaction between IT and workforce management. In A. Mesquita (Ed.), *Human interaction with technology for working, communicating, and learning: Advancements* (pp. 236–251). Hershey, PA: IGI Global. doi:10.4018/978-1-61350-465-9.ch015

Olaniran, B. A., Burley, H., & Chang, M. (2013). Social issues and web 2.0: A closer look at culture in e-learning. In *Digital literacy: Concepts, methodologies, tools, and applications* (pp. 1456–1471). Hershey, PA: IGI Global. doi:10.4018/978-1-4666-1852-7.ch077

Olaniran, B. A., Rodriguez, N., & Williams, I. M. (2012). Social information processing theory (SIPT): A cultural perspective for international online communication environments. In K. St. Amant & S. Kelsey (Eds.), *Computer-mediated communication across cultures: International interactions in online environments* (pp. 45–65). Hershey, PA: IGI Global. doi:10.4018/978-1-60960-833-0.ch004

Ololube, N. P., Amaele, S., Kpolovie, P. J., & Egbezor, D. E. (2013). The issues of digital natives and tourists: Empirical investigation of the level of IT/IS usage between university students and faculty members in a developing economy. In *Digital literacy: Concepts, methodologies, tools, and applications* (pp. 1384–1401). Hershey, PA: IGI Global. doi:10.4018/978-1-4666-1852-7.ch073

Orange, E. (2013). Understanding the human-machine interface in a time of change. In R. Luppicini (Ed.), *Handbook of research on technoself: Identity in a technological society* (pp. 703–719). Hershey, PA: IGI Global. doi:10.4018/978-1-4666-2211-1.ch036

Ouwehand, K., van Gog, T., & Paas, F. (2013). The use of gesturing to facilitate older adults' learning from computer-based dynamic visualizations. In R. Zheng, R. Hill, & M. Gardner (Eds.), *Engaging older adults with modern technology: Internet use and information access needs* (pp. 33–58). Hershey, PA: IGI Global. doi:10.4018/978-1-4666-1966-1.ch003

Owen, H. (2013). Intelligent m-learning frameworks: Information and communication technology applied in a laptop environment. In *Digital literacy: Concepts, methodologies, tools, and applications* (pp. 414–436). Hershey, PA: IGI Global. doi:10.4018/978-1-4666-1852-7.ch022

Oye, N. D., Iahad, N. A., & Rahim, N. Z. (2013). An application of the UTAUT model for understanding acceptance and use of ICT by Nigerian university academicians. In S. Chhabra (Ed.), *ICT influences on human development, interaction, and collaboration* (pp. 214–229). Hershey, PA: IGI Global. doi:10.4018/978-1-4666-1957-9.ch013

Özpinar, A., & Yavuz, E. (2013). E-commerce training with virtual commerce simulation. In S. Sharma (Ed.), *Adoption of virtual technologies for business, educational, and governmental advancements* (pp. 37–43). Hershey, PA: IGI Global. doi:10.4018/978-1-4666-2053-7.ch003

Papadopoulos, F., Dautenhahn, K., & Ho, W. C. (2013). Behavioral analysis of human-human remote social interaction mediated by an interactive robot in a cooperative game scenario. In R. Luppicini (Ed.), *Handbook of research on technoself: Identity in a technological society* (pp. 637–665). Hershey, PA: IGI Global. doi:10.4018/978-1-4666-2211-1.ch033

Paravati, G., Sanna, A., Lamberti, F., & Ciminiera, L. (2012). On quality of experience in remote visualization on mobile devices. In J. Lumsden (Ed.), *Social and organizational impacts of emerging mobile devices: Evaluating use* (pp. 1–19). Hershey, PA: IGI Global. doi:10.4018/978-1-4666-0194-9.ch001

Parker, D. (2013). Implementing the professional development program. In S. Wang & T. Hartsell (Eds.), *Technology integration and foundations for effective leadership* (pp. 190–205). Hershey, PA: IGI Global. doi:10.4018/978-1-4666-2656-0.ch011

Pee, L., Kankanhalli, A., & Show, V. O. (2013). ICT for digital inclusion: A study of public internet kiosks in Mauritius. In *Digital literacy: Concepts, methodologies, tools, and applications* (pp. 477–501). Hershey, PA: IGI Global. doi:10.4018/978-1-4666-1852-7.ch025

Peevers, G., Douglas, G., & Jack, M. A. (2013). Multimedia technology in the financial services sector: Customer satisfaction with alternatives to face-to-face interaction in mortgage sales. In A. Mesquita (Ed.), *User perception and influencing factors of technology in everyday life* (pp. 92–106). Hershey, PA: IGI Global. doi:10.4018/978-1-4666-1954-8.ch008

Peevers, G., Douglas, G., Jack, M. A., & Marshall, D. (2013). A usability comparison of SMS and IVR as digital banking channels. In A. Mesquita (Ed.), *User perception and influencing factors of technology in everyday life* (pp. 76–91). Hershey, PA: IGI Global. doi:10.4018/978-1-4666-1954-8.ch007

Pereira da Silva, H., & Loureiro da Silva, L. D. (2013). Digital inclusion and electronic government: Looking for convergence in the decade 1997-2008. In *Digital literacy: Concepts, methodologies, tools, and applications* (pp. 1192–1218). Hershey, PA: IGI Global. doi:10.4018/978-1-4666-1852-7.ch062

Pettitt, M., & Burnett, G. (2012). Visual demand evaluation methods for in-vehicle interfaces. In J. Lumsden (Ed.), *Social and organizational impacts of emerging mobile devices: Evaluating use* (pp. 265–278). Hershey, PA: IGI Global. doi:10.4018/978-1-4666-0194-9.ch015

Phipps, L. C., Wise, A., & Amundsen, C. (2013). The university in transition: Reconsidering faculty roles and expertise in a web 2.0 world. In T. Takševa (Ed.), *Social software and the evolution of user expertise: Future trends in knowledge creation and dissemination* (pp. 93–111). Hershey, PA: IGI Global. doi:10.4018/978-1-4666-2178-7.ch006

Polifroni, J., Kiss, I., & Seneff, S. (2013). Speech for content creation. In J. Lumsden (Ed.), *Developments in technologies for human-centric mobile computing and applications* (pp. 114–129). Hershey, PA: IGI Global. doi:10.4018/978-1-4666-2068-1.ch007

Popescu, E., & Badica, C. (2013). Creating a personalized artificial intelligence course: WELSA case study. In J. Wang (Ed.), *Information systems and modern society: Social change and global development* (pp. 31–48). Hershey, PA: IGI Global. doi:10.4018/978-1-4666-2922-6.ch003

Prakash, N. (2012). ICT and women empowerment in a rural setting in India. In R. Pande & T. Van der Weide (Eds.), *Globalization, technology diffusion and gender disparity: Social impacts of ICTs* (pp. 15–24). Hershey, PA: IGI Global. doi:10.4018/978-1-4666-0020-1.ch002

Prescott, J., & Bogg, J. (2013). Self, career, and gender issues: A complex interplay of internal/external factors. In *Gendered occupational differences in science, engineering, and technology careers* (pp. 79–111). Hershey, PA: IGI Global. doi:10.4018/978-1-4666-2107-7.ch004

Prescott, J., & Bogg, J. (2013). Stereotype, attitudes, and identity: Gendered expectations and behaviors. In *Gendered occupational differences in science, engineering, and technology careers* (pp. 112–135). Hershey, PA: IGI Global. doi:10.4018/978-1-4666-2107-7.ch005

Prescott, J., & Bogg, J. (2013). The computer games industry: New industry, same old issues. In *Gendered occupational differences in science, engineering, and technology careers* (pp. 64–77). Hershey, PA: IGI Global. doi:10.4018/978-1-4666-2107-7.ch003

Pure, R. A., Markov, A. R., Mangus, J. M., Metzger, M. J., Flanagin, A. J., & Hartsell, E. H. (2013). Understanding and evaluating source expertise in an evolving media environment. In T. Takševa (Ed.), *Social software and the evolution of user expertise: Future trends in knowledge creation and dissemination* (pp. 37–51). Hershey, PA: IGI Global. doi:10.4018/978-1-4666-2178-7.ch003

Quesenberry, J. L. (2012). Re-examining the career anchor model: An investigation of career values and motivations among women in the information technology profession. In R. Pande & T. Van der Weide (Eds.), *Globalization, technology diffusion and gender disparity: Social impacts of ICTs* (pp. 169–183). Hershey, PA: IGI Global. doi:10.4018/978-1-4666-0020-1.ch015

Quinlan, A., Quinlan, E., & Nelson, D. (2013). Performing actor-network theory in the post-secondary classroom. In A. Tatnall (Ed.), *Social and professional applications of actor-network theory for technology development* (pp. 56–66). Hershey, PA: IGI Global. doi:10.4018/978-1-4666-2166-4.ch005

Radu, R. G. (2012). From drift to draft: International institutional responses to the global digital divide. In R. Pande & T. Van der Weide (Eds.), *Globalization, technology diffusion and gender disparity: Social impacts of ICTs* (pp. 83–94). Hershey, PA: IGI Global. doi:10.4018/978-1-4666-0020-1.ch007

Raeth, P., & Smolnik, S. (2013). Towards a model of employee weblog usage: A process-oriented analysis of antecedents and consequences. In B. Medlin (Ed.), *Integrations of technology utilization and social dynamics in organizations* (pp. 125–141). Hershey, PA: IGI Global. doi:10.4018/978-1-4666-1948-7.ch008

Ragusa, A. T., & Steinke, E. (2012). Studying locally, interacting globally: Demographic change and international students in Australian higher education. In K. St.Amant & S. Kelsey (Eds.), *Computer-mediated communication across cultures: International interactions in online environments* (pp. 344–368). Hershey, PA: IGI Global. doi:10.4018/978-1-60960-833-0.ch022

Rahman, H., & Ramos, I. (2013). Empowerment of SMEs through open innovation strategies: Life cycle of technology management. In S. Chhabra (Ed.), *ICT influences on human development, interaction, and collaboration* (pp. 185–202). Hershey, PA: IGI Global. doi:10.4018/978-1-4666-1957-9.ch011

Raihan, A. (2012). Public access ICT in Bangladesh. In R. Gomez (Ed.), *Libraries, telecentres, cybercafes and public access to ICT: International comparisons* (pp. 249–266). Hershey, PA: IGI Global. doi:10.4018/978-1-60960-771-5.ch019

Rajesh, M. N. (2012). Virtual tourism as a new form of oppression against women. In R. Pande & T. Van der Weide (Eds.), *Globalization, technology diffusion and gender disparity: Social impacts of ICTs* (pp. 200–209). Hershey, PA: IGI Global. doi:10.4018/978-1-4666-0020-1.ch017

Ratan, R. (2013). Self-presence, explicated: Body, emotion, and identity extension into the virtual self. In R. Luppicini (Ed.), *Handbook of research on technoself: Identity in a technological society* (pp. 322–336). Hershey, PA: IGI Global. doi:10.4018/978-1-4666-2211-1.ch018

Reeves, T. C., & Herrington, J. (2013). Authentic tasks: The key to harnessing the drive to learn in members of "generation me". In *Digital literacy: Concepts, methodologies, tools, and applications* (pp. 1240–1255). Hershey, PA: IGI Global. doi:10.4018/978-1-4666-1852-7.ch064

Resta, P., & Kalk, D. (2013). An ecological approach to instructional design: The learning synergy of interaction and context. In *Digital literacy: Concepts, methodologies, tools, and applications* (pp. 872–889). Hershey, PA: IGI Global. doi:10.4018/978-1-4666-1852-7.ch045

Ribeiro, J. C., & Silva, T. (2013). Self, self-presentation, and the use of social applications in digital environments. In R. Luppicini (Ed.), *Handbook of research on technoself: Identity in a technological society* (pp. 439–455). Hershey, PA: IGI Global. doi:10.4018/978-1-4666-2211-1.ch024

Rienties, B., Tempelaar, D., Pinckaers, M., Giesbers, B., & Lichel, L. (2012). The diverging effects of social network sites on receiving job information for students and professionals. In E. Coakes (Ed.), *Technological change and societal growth: Analyzing the future* (pp. 202–217). Hershey, PA: IGI Global. doi:10.4018/978-1-4666-0200-7.ch013

Rimpiläinen, S. (2013). Knowledge in networks: Knowing in transactions? In A. Tatnall (Ed.), *Social and professional applications of actor-network theory for technology development* (pp. 45–55). Hershey, PA: IGI Global. doi:10.4018/978-1-4666-2166-4.ch004

Rive, P., & Thomassen, A. (2012). International collaboration and design innovation in virtual worlds: Lessons from Second Life. In K. St.Amant & S. Kelsey (Eds.), *Computer-mediated communication across cultures: International interactions in online environments* (pp. 429–448). Hershey, PA: IGI Global. doi:10.4018/978-1-60960-833-0.ch027

Rössler, T. (2012). Electronic voting using identity domain separation and hardware security modules. In S. Sharma (Ed.), *E-adoption and technologies for empowering developing countries: Global advances* (pp. 100–114). Hershey, PA: IGI Global. doi:10.4018/978-1-4666-0041-6.ch008

Rothblatt, M. (2013). Mindclone technoselves: Multi-substrate legal identities, cyber-psychology, and biocyberethics. In R. Luppicini (Ed.), *Handbook of research on technoself: Identity in a technological society* (pp. 105–122). Hershey, PA: IGI Global. doi:10.4018/978-1-4666-2211-1.ch006

Rozengardt, A., & Finquelievich, S. (2012). Public access ICT in Argentina. In R. Gomez (Ed.), *Libraries, telecentres, cybercafes and public access to ICT: International comparisons* (pp. 114–133). Hershey, PA: IGI Global. doi:10.4018/978-1-60960-771-5.ch011

Sabone, M. B., Mogobe, K. D., & Sabone, T. G. (2012). ICTS and their role in health promotion: A preliminary situation analysis in selected Botswana rural communities. In R. Lekoko & L. Semali (Eds.), *Cases on developing countries and ICT integration: Rural community development* (pp. 1–15). Hershey, PA: IGI Global. doi:10.4018/978-1-60960-117-1.ch001

Saleh, A. I. (2013). A novel strategy for managing user's locations in PCS networks based on a novel hot spots topology. In S. Chhabra (Ed.), *ICT influences on human development, interaction, and collaboration* (pp. 43–77). Hershey, PA: IGI Global. doi:10.4018/978-1-4666-1957-9.ch003

Samiei, A., & Laitsch, D. A. (2012). The concerns of elementary educators with the diffusion of information and communication technology. In S. Chhabra (Ed.), *ICTs for advancing rural communities and human development: Addressing the digital divide* (pp. 193–207). Hershey, PA: IGI Global. doi:10.4018/978-1-4666-0047-8.ch013

Sandulli, F. D. (2013). Solving the paradoxes of the information technology revolution: Productivity and inequality. In *Digital literacy: Concepts, methodologies, tools, and applications* (pp. 1440–1455). Hershey, PA: IGI Global. doi:10.4018/978-1-4666-1852-7.ch076

Santana, P. C., Ahumada, A. C., & Magaña, M. A. (2013). Usability testing of an education management information system: The case of the University of Colima. In M. Garcia-Ruiz (Ed.), *Cases on usability engineering: Design and development of digital products* (pp. 80–93). Hershey, PA: IGI Global. doi:10.4018/978-1-4666-4046-7.ch004

Schäler, R. (2012). Information sharing across languages. In K. St.Amant & S. Kelsey (Eds.), *Computer-mediated communication across cultures: International interactions in online environments* (pp. 215–234). Hershey, PA: IGI Global. doi:10.4018/978-1-60960-833-0.ch015

Schleicher, R., Shirazi, A. S., Rohs, M., Kratz, S., & Schmidt, A. (2013). WorldCupinion: Experiences with an android app for real-time opinion sharing during soccer world cup games. In J. Lumsden (Ed.), *Developments in technologies for human-centric mobile computing and applications* (pp. 240–257). Hershey, PA: IGI Global. doi:10.4018/978-1-4666-2068-1.ch015

Schott, G., & Selwyn, N. (2013). Game literacy: Assessing its value for both classification and public perceptions of games in a New Zealand context. In *Digital literacy: Concepts, methodologies, tools, and applications* (pp. 347–363). Hershey, PA: IGI Global. doi:10.4018/978-1-4666-1852-7.ch018

Schulmeister, R. (2013). Students, internet, elearning and web 2.0. In *Digital literacy: Concepts, methodologies, tools, and applications* (pp. 1720–1742). Hershey, PA: IGI Global. doi:10.4018/978-1-4666-1852-7.ch090

Scolari, C. A., Romaní, C. C., & Kuklinski, H. P. (2013). Should we take disintermediation in higher education seriously? Expertise, knowledge brokering, and knowledge translation in the age of disintermediation. In T. Takševa (Ed.), *Social software and the evolution of user expertise: Future trends in knowledge creation and dissemination* (pp. 72–92). Hershey, PA: IGI Global. doi:10.4018/978-1-4666-2178-7.ch005

Segal, J., & Morris, C. (2013). Scientific end-user developers and barriers to user/customer engagement. In A. Dwivedi & S. Clarke (Eds.), *Innovative strategies and approaches for end-user computing advancements* (pp. 333–346). Hershey, PA: IGI Global. doi:10.4018/978-1-4666-2059-9.ch018

Selg, H. (2013). Swedish students online: An inquiry into differing cultures on the internet. In *Digital literacy: Concepts, methodologies, tools, and applications* (pp. 735–756). Hershey, PA: IGI Global. doi:10.4018/978-1-4666-1852-7.ch038

Shen, J., & Eder, L. B. (2013). An examination of factors associated with user acceptance of social shopping websites. In A. Mesquita (Ed.), *User perception and influencing factors of technology in everyday life* (pp. 28–45). Hershey, PA: IGI Global. doi:10.4018/978-1-4666-1954-8.ch004

Sidhu, M. S., & Kang, L. C. (2012). New trends and futuristic information communication technologies for engineering education. In S. Chhabra (Ed.), *ICTs for advancing rural communities and human development: Addressing the digital divide* (pp. 251–262). Hershey, PA: IGI Global. doi:10.4018/978-1-4666-0047-8.ch017

Sierra, W., & Eyman, D. (2013). "I rolled the dice with trade chat and this is what i got": Demonstrating context-dependent credibility in virtual worlds. In M. Folk & S. Apostel (Eds.), *Online credibility and digital ethos: Evaluating computer-mediated communication* (pp. 332–352). Hershey, PA: IGI Global. doi:10.4018/978-1-4666-2663-8.ch019

Sintoris, C., Stoica, A., Papadimitriou, I., Yiannoutsou, N., Komis, V., & Avouris, N. (2012). MuseumScrabble: Design of a mobile game for children's interaction with a digitally augmented cultural space. In J. Lumsden (Ed.), *Social and organizational impacts of emerging mobile devices: Evaluating use* (pp. 124–142). Hershey, PA: IGI Global. doi:10.4018/978-1-4666-0194-9.ch007

Slegers, K., & van Boxtel, M. P. (2013). Actual use of computers and the internet by older adults: Potential benefits and risks. In R. Zheng, R. Hill, & M. Gardner (Eds.), *Engaging older adults with modern technology: Internet use and information access needs* (pp. 161–190). Hershey, PA: IGI Global. doi:10.4018/978-1-4666-1966-1.ch009

Smith, P. (2012). Affective factors for successful knowledge management. In E. Coakes (Ed.), *Technological change and societal growth: Analyzing the future* (pp. 79–87). Hershey, PA: IGI Global. doi:10.4018/978-1-4666-0200-7.ch005

Söderström, S. (2013). Assistive ICT and young disabled persons: Opportunities and obstacles in identity negotiations. In R. Luppicini (Ed.), *Handbook of research on technoself: Identity in a technological society* (pp. 337–358). Hershey, PA: IGI Global. doi:10.4018/978-1-4666-2211-1.ch019

Sone, Y. (2013). Robot double: Hiroshi Ishiguro's reflexive machines. In R. Luppicini (Ed.), *Handbook of research on technoself: Identity in a technological society* (pp. 680–702). Hershey, PA: IGI Global. doi:10.4018/978-1-4666-2211-1.ch035

Sørensen, B. H., & Levinsen, K. T. (2013). School in the knowledge society: A local global school. In *Digital literacy: Concepts, methodologies, tools, and applications* (pp. 959–975). Hershey, PA: IGI Global. doi:10.4018/978-1-4666-1852-7.ch049

Spence, E. (2012). Luciano Floridi's metaphysical theory of information ethics: A critical appraisal and an alternative neo-Gewirthian information ethics. In A. Mesquita (Ed.), *Human interaction with technology for working, communicating, and learning: Advancements* (pp. 134–148). Hershey, PA: IGI Global. doi:10.4018/978-1-61350-465-9.ch009

Stam, K., Guzman, I. R., & Thoryk, D. (2012). Studying online communication features on international and cross-cultural web pages using websphere analysis methodology. In K. St.Amant & S. Kelsey (Eds.), *Computer-mediated communication across cultures: International interactions in online environments* (pp. 187–199). Hershey, PA: IGI Global. doi:10.4018/978-1-60960-833-0.ch013

Stephens, R. T. (2013). Usability impact analysis of collaborative environments. In M. Garcia-Ruiz (Ed.), *Cases on usability engineering: Design and development of digital products* (pp. 94–116). Hershey, PA: IGI Global. doi:10.4018/978-1-4666-4046-7.ch005

Sterbini, A., & Temperini, M. (2013). SocialX: An improved reputation based support to social collaborative learning through exercise sharing and project teamwork. In J. Wang (Ed.), *Information systems and modern society: Social change and global development* (pp. 66–85). Hershey, PA: IGI Global. doi:10.4018/978-1-4666-2922-6.ch005

Stevens, G., Pipek, V., & Wulf, V. (2012). Appropriation infrastructure: Mediating appropriation and production work. In A. Dwivedi & S. Clarke (Eds.), *End-user computing, development, and software engineering: New challenges* (pp. 254–278). Hershey, PA: IGI Global. doi:10.4018/978-1-4666-0140-6.ch012

Strömberg-Jakka, M. (2013). Social assistance via the internet: The case of Finland in the European context. In J. Lannon & E. Halpin (Eds.), *Human rights and information communication technologies: Trends and consequences of use* (pp. 177–195). Hershey, PA: IGI Global. doi:10.4018/978-1-4666-1918-0.ch011

Subrahmanyam, K., & Manago, A. (2012). The children's digital media center @ Los Angeles. In Z. Yan (Ed.), *Encyclopedia of cyber behavior* (pp. 64–76). Hershey, PA: IGI Global. doi:10.4018/978-1-4666-0315-8.ch005

Suki, N. M. (2013). Modelling factors influencing early adopters' purchase intention towards online music. In A. Mesquita (Ed.), *User perception and influencing factors of technology in everyday life* (pp. 298–314). Hershey, PA: IGI Global. doi:10.4018/978-1-4666-1954-8.ch020

Sweet, C. (2013). Wikipedia's success and the rise of the amateur-expert. In T. Takševa (Ed.), *Social software and the evolution of user expertise: Future trends in knowledge creation and dissemination* (pp. 13–36). Hershey, PA: IGI Global. doi:10.4018/978-1-4666-2178-7.ch002

Swindell, R., Grimbeek, P., & Heffernan, J. (2013). U3A online and successful aging: A smart way to help bridge the grey digital divide. In *Digital literacy: Concepts, methodologies, tools, and applications* (pp. 1640–1657). Hershey, PA: IGI Global. doi:10.4018/978-1-4666-1852-7.ch086

Sylaiou, S., White, M., & Liarokapis, F. (2013). Digital heritage systems: The ARCO evaluation. In M. Garcia-Ruiz (Ed.), *Cases on usability engineering: Design and development of digital products* (pp. 321–354). Hershey, PA: IGI Global. doi:10.4018/978-1-4666-4046-7.ch014

Tai, Z., & Zhang, Y. (2013). Online identity formation and digital ethos building in the Chinese blogosphere. In M. Folk & S. Apostel (Eds.), *Online credibility and digital ethos: Evaluating computer-mediated communication* (pp. 269–284). Hershey, PA: IGI Global. doi:10.4018/978-1-4666-2663-8.ch015

Taiwo, R. (2012). Discursive manipulation strategies in virtual scams in global contexts. In K. St.Amant & S. Kelsey (Eds.), *Computer-mediated communication across cultures: International interactions in online environments* (pp. 143–154). Hershey, PA: IGI Global. doi:10.4018/978-1-60960-833-0.ch010

Tchouakeu, L. N., Maldonado, E., Zhao, K., Robinson, H., Maitland, C., & Tapia, A. (2013). Exploring barriers to coordination between humanitarian NGOs: A comparative case study of two NGO's information technology coordination bodies. In J. Wang (Ed.), *Information systems and modern society: Social change and global development* (pp. 87–112). Hershey, PA: IGI Global. doi:10.4018/978-1-4666-2922-6.ch006

Tella, A., Mutula, S. M., Mutshewa, A., & Totolo, A. (2012). An evaluation of WebCT course content management system at the university of Botswana. In S. Sharma (Ed.), *E-adoption and technologies for empowering developing countries: Global advances* (pp. 205–232). Hershey, PA: IGI Global. doi:10.4018/978-1-4666-0041-6.ch015

Thatcher, A., & Ndabeni, M. (2013). A psychological model to understand e-adoption in the context of the digital divide. In *Digital literacy: Concepts, methodologies, tools, and applications* (pp. 1402–1424). Hershey, PA: IGI Global. doi:10.4018/978-1-4666-1852-7.ch074

Todd, P. M., Rogers, Y., & Payne, S. J. (2013). Nudging the trolley in the supermarket: How to deliver the right information to shoppers. In J. Lumsden (Ed.), *Developments in technologies for human-centric mobile computing and applications* (pp. 99–113). Hershey, PA: IGI Global. doi:10.4018/978-1-4666-2068-1.ch006

Tojib, D. R., & Sugianto, L. (2013). Construct validity assessment in IS research: Methods and case example of user satisfaction scale. In A. Dwivedi & S. Clarke (Eds.), *Innovative strategies and approaches for end-user computing advancements* (pp. 134–159). Hershey, PA: IGI Global. doi:10.4018/978-1-4666-2059-9.ch008

Tregeagle, S. (2012). Participation in child welfare services through information and communication technologies. In S. Chhabra (Ed.), *ICTs for advancing rural communities and human development: Addressing the digital divide* (pp. 73–90). Hershey, PA: IGI Global. doi:10.4018/978-1-4666-0047-8.ch006

Truong, Y. (2013). Antecedents of consumer acceptance of mobile television advertising. In A. Mesquita (Ed.), *User perception and influencing factors of technology in everyday life* (pp. 128–141). Hershey, PA: IGI Global. doi:10.4018/978-1-4666-1954-8.ch010

Truyen, F., & Buekens, F. (2013). Professional ICT knowledge, epistemic standards, and social epistemology. In T. Takševa (Ed.), *Social software and the evolution of user expertise: Future trends in knowledge creation and dissemination* (pp. 274–294). Hershey, PA: IGI Global. doi:10.4018/978-1-4666-2178-7.ch016

Tsai, C. (2013). How much can computers and internet help? A long-term study of web-mediated problem-based learning and self-regulated learning. In A. Mesquita (Ed.), *User perception and influencing factors of technology in everyday life* (pp. 248–264). Hershey, PA: IGI Global. doi:10.4018/978-1-4666-1954-8.ch017

Tsai, C., & Shen, P. (2013). Improving students' computing skills and attitudes toward learning via web-mediated self-regulated learning with feedback in an online problem-solving environment. In S. Sharma (Ed.), *Adoption of virtual technologies for business, educational, and governmental advancements* (pp. 162–179). Hershey, PA: IGI Global. doi:10.4018/978-1-4666-2053-7.ch014

Unsal, F., Komaromi, K., & Erickson, G. S. (2013). Trust in e-commerce: Social networks vs. institutional credibility. In S. Sharma (Ed.), *Adoption of virtual technologies for business, educational, and governmental advancements* (pp. 44–57). Hershey, PA: IGI Global. doi:10.4018/978-1-4666-2053-7.ch004 van de Sande, C. (2012). Free, open, online help forums: convenience, connection, control, comfort, and communication. In E. Coakes (Ed.), *Technological change and societal growth: Analyzing the future* (pp. 162–179). Hershey, PA: IGI Global. doi:10.4018/978-1-4666-0200-7.ch011

van der Weide, T. (2012). A digital (r)evolution to the information age. In R. Pande & T. Van der Weide (Eds.), *Globalization, technology diffusion and gender disparity: Social impacts of ICTs* (pp. 1–14). Hershey, PA: IGI Global. doi:10.4018/978-1-4666-0020-1.ch001

van Dijck, J. (2013). Google scholar as the co-producer of scholarly knowledge. In T. Takševa (Ed.), *Social software and the evolution of user expertise: Future trends in knowledge creation and dissemination* (pp. 130–146). Hershey, PA: IGI Global. doi:10.4018/978-1-4666-2178-7.ch008

Varga-Atkins, T., Prescott, D., & Dangerfield, P. (2012). Cyber behavior with wikis. In Z. Yan (Ed.), *Encyclopedia of cyber behavior* (pp. 164–177). Hershey, PA: IGI Global. doi:10.4018/978-1-4666-0315-8.ch014

Veil, S. R. (2012). Adoption barriers in a high-risk agricultural environment. In A. Mesquita (Ed.), *Human interaction with technology for working, communicating, and learning: Advancements* (pp. 31–47). Hershey, PA: IGI Global. doi:10.4018/978-1-61350-465-9.ch003

Walczak, S., & Mann, R. (2012). Utilization and perceived benefit for diverse users of communities of practice in a healthcare organization. In A. Dwivedi & S. Clarke (Eds.), *End-user computing, development, and software engineering: New challenges* (pp. 25–53). Hershey, PA: IGI Global. doi:10.4018/978-1-4666-0140-6.ch002

Wallgren, L. G., Leijon, S., & Andersson, K. M. (2013). IT managers' narratives on subordinates' motivation at work: A case study. In A. Mesquita (Ed.), *User perception and influencing factors of technology in everyday life* (pp. 282–297). Hershey, PA: IGI Global. doi:10.4018/978-1-4666-1954-8.ch019

Walton, A. L., DeVaney, S. A., & Sandall, D. L. (2013). Graduate students' perceptions of privacy and closed circuit television systems in public settings. In A. Mesquita (Ed.), *User perception and influencing factors of technology in everyday life* (pp. 107–127). Hershey, PA: IGI Global. doi:10.4018/978-1-4666-1954-8.ch009

Wang, J., Doll, W. J., & Deng, X. (2012). A model of system re-configurability and pedagogical usability in an e-learning context: A faculty perspective. In A. Dwivedi & S. Clarke (Eds.), *End-user computing, development, and software engineering: New challenges* (pp. 168–184). Hershey, PA: IGI Global. doi:10.4018/978-1-4666-0140-6.ch008

Wang, J., Zhou, B., & Hsu, J. (2012). Assessment and contrast of the effects of information and communication technology. In S. Chhabra (Ed.), *ICTs for advancing rural communities and human development: Addressing the digital divide* (pp. 15–32). Hershey, PA: IGI Global. doi:10.4018/978-1-4666-0047-8.ch002

Wang, Y., Wang, X., Qian, Y., Luo, H., Ge, F., Yang, Y., & Xia, Y. (2013). Residential load pattern analysis for smart grid applications based on audio feature EEUPC. In T. Gao (Ed.), *Global applications of pervasive and ubiquitous computing* (pp. 107–121). Hershey, PA: IGI Global. doi:10.4018/978-1-4666-2645-4.ch013

Wang, Y. K., & Datta, P. (2012). Investigating technology commitment in instant messaging application users. In A. Dwivedi & S. Clarke (Eds.), *End-user computing, development, and software engineering: New challenges* (pp. 227–252). Hershey, PA: IGI Global. doi:10.4018/978-1-4666-0140-6.ch011

Was, C. A., & Woltz, D. J. (2013). Implicit memory and aging: Adapting technology to utilize preserved memory functions. In R. Zheng, R. Hill, & M. Gardner (Eds.), *Engaging older adults with modern technology: Internet use and information access needs* (pp. 1–19). Hershey, PA: IGI Global. doi:10.4018/978-1-4666-1966-1.ch001

Watson, J. M., Lambert, A. E., Cooper, J. M., Boyle, I. V., & Strayer, D. L. (2013). On attentional control and the aging driver. In R. Zheng, R. Hill, & M. Gardner (Eds.), *Engaging older adults with modern technology: Internet use and information access needs* (pp. 20–32). Hershey, PA: IGI Global. doi:10.4018/978-1-4666-1966-1.ch002

Watson, S. L., & Mulvihill, T. (2013). Exploring the notion of 'technology as a public good': Emerging characteristics and trends of the digital divide in East Asian education. In *Digital literacy: Concepts, methodologies, tools, and applications* (pp. 1472–1488). Hershey, PA: IGI Global. doi:10.4018/978-1-4666-1852-7.ch078

Watts, C. (2013). Connection, fragmentation, and intentionality: Social software and the changing nature of expertise. In T. Takševa (Ed.), *Social software and the evolution of user expertise: Future trends in knowledge creation and dissemination* (pp. 52–70). Hershey, PA: IGI Global. doi:10.4018/978-1-4666-2178-7.ch004

Watts, J. D., Kanyasone, V., & Vongkhamsao, V. (2012). Pathways to participatory landscape governance in northern Laos: The role of information and communication technologies. In S. Chhabra (Ed.), *ICTs for advancing rural communities and human development: Addressing the digital divide* (pp. 143–157). Hershey, PA: IGI Global. doi:10.4018/978-1-4666-0047-8.ch010

Weber, R. H. (2013). ICT policies favouring human rights. In J. Lannon & E. Halpin (Eds.), *Human rights and information communication technologies: Trends and consequences of use* (pp. 21–35). Hershey, PA: IGI Global. doi:10.4018/978-1-4666-1918-0.ch002

Wei, Z., & Kramarae, C. (2012). Women, big ideas, and social networking technologies: Hidden assumptions. In R. Pande & T. Van der Weide (Eds.), *Globalization, technology diffusion and gender disparity: Social impacts of ICTs* (pp. 70–82). Hershey, PA: IGI Global. doi:10.4018/978-1-4666-0020-1.ch006

Whiteside, A. L., & Dikkers, A. G. (2012). Maximizing multicultural online learning experiences with the social presence model, course examples, and specific strategies. In K. St.Amant & S. Kelsey (Eds.), *Computer-mediated communication across cultures: International interactions in online environments* (pp. 395–413). Hershey, PA: IGI Global. doi:10.4018/978-1-60960-833-0.ch025

Widyanto, L., & Griffiths, M. (2013). An empirical study of problematic internet use and self-esteem. In R. Zheng (Ed.), *Evolving psychological and educational perspectives on cyber behavior* (pp. 82–95). Hershey, PA: IGI Global. doi:10.4018/978-1-4666-1858-9.ch006

Wiebelhaus-Brahm, E. (2013). Truth-seeking at a distance: Engaging diaspora populations in transitional justice processes. In J. Lannon & E. Halpin (Eds.), *Human rights and information communication technologies: Trends and consequences of use* (pp. 72–85). Hershey, PA: IGI Global. doi:10.4018/978-1-4666-1918-0.ch005

Wilhelmson, L., Johansson, P., & Döös, M. (2013). Bridging boundaries: Middle managers' pedagogic interventions as technology leaders. In S. Wang & T. Hartsell (Eds.), *Technology integration and foundations for effective leadership* (pp. 278–292). Hershey, PA: IGI Global. doi:10.4018/978-1-4666-2656-0.ch016

Wilson, C., & Dunn, A. (2013). Contingency and hybridity in the study of digital advocacy networks: Implications of the Egyptian protest movement. In J. Lannon & E. Halpin (Eds.), *Human rights and information communication technologies: Trends and consequences of use* (pp. 100–121). Hershey, PA: IGI Global. doi:10.4018/978-1-4666-1918-0.ch007

Wilson, S., & Haslam, N. (2013). Reasoning about human enhancement: Towards a folk psychological model of human nature and human identity. In R. Luppicini (Ed.), *Handbook of research on technoself: Identity in a technological society* (pp. 175–188). Hershey, PA: IGI Global. doi:10.4018/978-1-4666-2211-1.ch010

Woodfield, R. (2012). Gender and employability patterns amongst UK ICT graduates: Investigating the leaky pipeline. In R. Pande & T. Van der Weide (Eds.), *Globalization, technology diffusion and gender disparity: Social impacts of ICTs* (pp. 184–199). Hershey, PA: IGI Global. doi:10.4018/978-1-4666-0020-1.ch016

Yan, Z., & Zheng, R. Z. (2013). Growing from childhood into adolescence: The science of cyber behavior. In R. Zheng (Ed.), *Evolving psychological and educational perspectives on cyber behavior* (pp. 1–14). Hershey, PA: IGI Global. doi:10.4018/978-1-4666-1858-9.ch001

Yang, P. (2012). Knowing through asynchronous time and space: A phenomenological study of cultural differences in online interaction. In K. St.Amant & S. Kelsey (Eds.), *Computer-mediated communication across cultures: International interactions in online environments* (pp. 108–122). Hershey, PA: IGI Global. doi:10.4018/978-1-60960-833-0.ch008

Yi, M. (2012). Minimizing cultural differences using ontology-based information retrieval system. In K. St.Amant & S. Kelsey (Eds.), *Computer-mediated communication across cultures: International interactions in online environments* (pp. 200–214). Hershey, PA: IGI Global. doi:10.4018/978-1-60960-833-0.ch014

Yocom, H. (2013). Implementation of the technology plan. In S. Wang & T. Hartsell (Eds.), *Technology integration and foundations for effective leadership* (pp. 102–123). Hershey, PA: IGI Global. doi:10.4018/978-1-4666-2656-0.ch007

Yoo, J., Hwang, W., Seok, H., Park, S. K., Kim, C., Choi, W. H., & Park, K. (2012). Cocktail: Exploiting bartenders' gestures for mobile interaction. In J. Lumsden (Ed.), *Social and organizational impacts of emerging mobile devices: Evaluating use* (pp. 186–199). Hershey, PA: IGI Global. doi:10.4018/978-1-4666-0194-9.ch011

Zaman, M., Simmers, C. A., & Anandarajan, M. (2013). Using an ethical framework to examine linkages between "going green" in research practices and information and communication technologies. In B. Medlin (Ed.), *Integrations of technology utilization and social dynamics in organizations* (pp. 243–262). Hershey, PA: IGI Global. doi:10.4018/978-1-4666-1948-7.ch015

Zhang, L., & Jones, M. C. (2013). A social capital perspective on IT professionals' work behavior and attitude. In A. Dwivedi & S. Clarke (Eds.), *Innovative strategies and approaches for end-user computing advancements* (pp. 160–174). Hershey, PA: IGI Global. doi:10.4018/978-1-4666-2059-9.ch009

Zheng, R. Z. (2013). Effective online learning for older people: A heuristic design approach. In R. Zheng, R. Hill, & M. Gardner (Eds.), *Engaging older adults with modern technology: Internet use and information access needs* (pp. 142–159). Hershey, PA: IGI Global. doi:10.4018/978-1-4666-1966-1.ch008

Zolfaghar, K., Khoshalhan, F., & Rabiei, M. (2012). User acceptance of location-based mobile advertising: An empirical study in Iran. In S. Sharma (Ed.), *E-adoption and technologies for empowering developing countries: Global advances* (pp. 59–71). Hershey, PA: IGI Global. doi:10.4018/978-1-4666-0041-6.ch005

Compilation of References

Accenture. (2016). *Accenture Technology Vision 2016: People First: The Primacy of People in a Digital Age*. Retrieved from http://goo.gl/JGfQvJ

Accenture. (2017). *Technology Trends 2017 – Accenture Tech Vision*. Retrieved March 31, 2017 from https://www.accenture.com/us-en/insight-disruptive-technology-trends-2017

Adams, N. B. (2007). Toward a model for knowledge development in virtual environments: Strategies for student ownership. *International Journal for Social Sciences*, *2*(2), 71–77.

Adobe. (2013). *Digital trends for 2013*. Retrieved 2016, from http://success.adobe.com/en/uk/programs/products/digitalmarketing/1301-28808-econsultancy-digital-trends-for-2013.html

Agarwal, N., Liu, H., Tang, L., & Yu, P. S. (2008, February). Identifying the influential bloggers in a community. In *Proceedings of the 2008 international conference on web search and data mining* (pp. 207 - 218). ACM. doi:10.1145/1341531.1341559

Ahmad, A. (2011). Rising of Social Network Websites in India Overview. *IJCSNS International Journal of Computer Science and Network Security*, *11*, 156.

Alade, I. A. (2004). Improving the status of vocational technical education: A route to National development. In D. F. Elaturoti & K. Babarinde (Eds.), *Teachers' Mandate on Education and Social Development in Nigeria* (pp. 95–105). Ibadan: Sterling-Horden Publishers.

Allen, J., & Velden, R. K. W. (2012). *Skills for the 21st century: Implications for education. Meteor*. Maastricht University School of Business and Economics.

Almarabeh, T., Mohammad, H., Yousef, R., & Majdalawi, Y. K. (2014). The university of Jordan e-learning platform: State, students' acceptance and challenges. *Journal of Software Engineering and Applications*, *7*(12), 99–107. doi:10.4236/jsea.2014.712087

Alonso, F., López, G., Manrique, D., & Viñes, J. M. (2005). An instructional model for web-based e-learning education with a blended learning process approach. *British Journal of Educational Technology*, *36*(2), 217–235. doi:10.1111/j.1467-8535.2005.00454.x

Alvesson, M., & Kärreman, D. (2001). Odd couple: Making sense of the curious concept of knowledge management. *Journal of Management Studies*, *38*(7), 995–1018. doi:10.1111/1467-6486.00269

American Society for Training and Development (ASTD). (2010). *State of industry report: Executive summary. Alexandria.* Alexandria: VAASTD Press.

Amodio, M. (2013, April 08). The Challenges of Workforce Management and How to Triumph Over Them. *Workforcemanagementtoday*. Retrieved from http://www.workforcemanagementtoday.com/articles/333395-challenges-workforce-management-how-triumph-over-them.html

Anderson, L. W. (2004). *Increasing teacher effectiveness* (2nd ed.). UNESCO: International Institute for Planning.

Ansari, A., & Mohd., A. (2009). Role of venture capital in spurring innovation and entrepreneurship. Asia - Pacific Business Review, (1), 114-119.

Aquino, J. (2016, August 24). Artificial Intelligence and the Future of Work. Retrieved July 21, 2017 from http://www.1to1media.com/technology/artificial-intelligence-and-future-work

Arabasz, P., Pirani, J., Pond, S., & Fawcett, D. (2003). *Supporting E-Learning in Higher Education*. Retrieved from http://net.educause.edu/ir/library/pdf/ers0303/rs/ers0303w.pdf

Ardito, C., Costabile, M. F., De Marsico, M., Lanzilotti, R., Levialdi, S., Roselli, T., & Rossano, V. (2006). An approach to usability evaluation of e-learning applications. *Universal Access in the Information Society*, *4*(3), 270–283. doi:10.1007/s10209-005-0008-6

Arias, F., Basualdo, L., & Heredia, V. (1989). Performance measurement. In F. Arias (Ed.), *Human Resources Administration (4th ed.)*. México: Trillas.

Arkorful, V., & Abaidoo, N. (2015). The role of e-learning, advantages and disadvantages of its adoption in higher education. *International Journal of Instructional Technology and Distance Learning*, *12*(1), 29–42.

Ashiyan, Z., & Salehi, H. (2016). Impact of WhatsApp on Learning and Retention of Collocation Knowledge among Iranian EFL Learners. *Advances in Language and Literary Studies*, *7*(5), 112–127.

Atiku, S. O., Chitakunye, P., & Fields, Z. (2014). Influence of organisational learning and human resource outcomes on commercial banks' performance in Nigeria. *J. Soc. Sci.*, *40*(1), 9–20.

Atiku, S. O., & Fields, Z. (2017). Multicultural Orientations for 21st Century Global Leadership. In N. Baporikar (Ed.), *Management Education for Global Leadership* (pp. 28–51). Hershey, USA: IGI Global. doi:10.4018/978-1-5225-1013-0.ch002

Bain, L., & Norris, S. (n. d.). Assessing the impact of applications of digital health records on Alzheimer's disease research.

Bakhtiyari, A., Barros, A., & Stewart, G. (2014). *Applying Enterprise Architecture to Business Networks*. Retrieved from http://eprints.qut.edu.au/73012/1/73012.pdf

Barnett, C. (2013). Top 10 Crowd funding Sites for Fundraising. *Forbes*. Retrieved from http://www.forbes.com/sites/chancebarnett/2013/05/08/top-10-crowdfunding-sites-for-fundraising/#3e9d76eb1cfb

Barrett, B. (2011). Developing and implementing strategies to enhance intellectual capital in the online learning environment. In *Proceedings of the 3rd European Conference on Intellectual Capital ECIC '11* (p. 62). Academic Conferences Limited.

Beal, V. (n. d.). *E-learning - Webopedia Definition*. Retrieved January 12, 2017 from http://www.webopedia.com/TERM/E/e_learning.html

Belausteguigoitia Rius, I. (2003). *Family businesses. Its dynamics, balance and consolidation*. México: McGraw - Hill.

Benavides, E., & Delgado, Y. L. (2004, May). Analysis of situational leadership and disposition towards tasks based on skill and ability. In *Memories of the VI International Congress of Ergonomics,* Universidad de Guanajuato.

Berge, Z. (2002). Obstacles to distance training and education in corporate organizations. *Journal of Workplace Learning*, *14*(5), 182–189. doi:10.1108/13665620210433873

Bhattacharya, M., Harold-Doty, D., & Garavan, T. (2014). The Organizational Context and Performance Implications of Human Capital Investment Variability. *Human Resource Development Quarterly*, *25*(1), 87–113. doi:10.1002/hrdq.21182

Biedrzycki, N. (2017). *Digitization is not a trend: Digitization is already our reality*. Retrieved March 31, 2017 from https://norbertbiedrzycki.pl/en/digital-journey

Blanc, B., & Figliozzi, M. (2016). Modeling the impacts of facility type, trip characteristics, and trip stressors on cyclists' comfort levels utilizing crowd sourced data. *Journal of the Transportation Research Board*, (2587), 100 - 108.

Blanchard, K., Zigarmi, P., Zigarmi, D., & Halsey, V. (2013). *Situational Leadership II*. The Ken Blanchar Companies. Retrieved February 12, 2016, from http://images.leadership.kenblanchard.com/Web/TheKenBlanchardCompanies/%7B26e2fdd1-daa7-4b13-be24-752d41d34a1b%7D_The-SLII-Experience-Sample-PW.pdf

Blanchard, K. (2007). *Leadership at the highest level. How to Create and Lead High Performance Organizations*. México: Editorial Norma.

Blewett, C., & Hugo, W. (2016). The learning affordance of a Facebook environment. In K. Pillay & C. Goodier (Eds.), *Teaching and Learning in the College of Law and Management Studies: Shared Approaches, Lesson and Good Practices* (pp. 46–63). Durban, South Africa: University of KwaZulu-Natal Publication.

Boorsma, B. (2017). *Digitization: Challenges & Opportunities. blogs@Cisco-Cisco Blogs*. Retrieved March 31, 2017 from http://blogs.cisco.com/government/digitization-challenges-opportunities

Bostman, R., & Rogers, R. (2010). *What's mine is yours: the rise of collaborative consumption*. New York: Harper Business.

Brown, L. (2017). *5 Dangers of Ransom ware everyone should know about*. Retrieved from https://www.shredit.com/en-us/blog/securing-your-information/february-2017/5-dangers-of-ransomware-everyone-should-know-about

Canadian Council on Learning. (2009). *State of e-learning and the science of instruction, proven guidelines for consumers and designers of multimedia learning* (3rd ed.). San Francisco: Pfeiffer.

Canals, J. (1994). *La internacionalización de la empresa*. Madrid: McGraw-Hill.

Cañigueral, A. (2015). Consumo colaborativo. *Leaners Magazine*, *2015*, 18-22.

Carbaugh, R. J. (2005). La globalización de la actividad económica. In R. J. Carbaugh (Ed.), Economía Internacional (pp. 4-8). Thompson.

Cassidy, J. (2006). Me media: How hanging out on the Internet became big Business. *The new Yorker*, *82*, 13-50.

Caudill, J. G. (2015). Employee Motivations for Workplace Learning and the Role of E-learning in the Workplace. *Internet Learning*, *4*(2), 37–48. doi:10.18278/il.4.2.3

CFE. (2012). *Electricity sector works and investments program 2012 - 2016*. México: CFE.

Chafkin, M. (2016, January). Can AirBnB unite the world? *Fast Company*. Retrieved from http://www.fastcompany.com/3054873/can-airbnb-unite-the-world

Chang, V. (2016). Review and discussion: E-learning for academia and industry. *International Journal of Information Management*, *36*(3), 476–485. doi:10.1016/j.ijinfomgt.2015.12.007

Chau, M., Wong, A., Wang, M., Lai, S., Chan, K. W., Li, T. M., & Sung, W. K. (2013). Using 3D virtual environments to facilitate students in constructivist learning. *Decision Support Systems*, *56*, 115–121. doi:10.1016/j.dss.2013.05.009

Chiavenato, I. (2002). *Human Talent Management*. Bogota, Colombia: Editorial McGraw-Hill.

Chiavenato, I. (2000). *Human resources management*. Bogotá, Colombia: Editorial Mc Graw –Hill.

Chimezie, A. (1988). The challenge of responsibilities: Confronting the revolution in workplace rights in modern organizations. *Employee Responsibilities and Rights Journal*, *1*(1), 5-23. doi:10.1007/bf01385449

Christensen, C., & Raynor, M. (2003). *The innovator's solution: creating and sustaining successful growth*. Boston: Harvard Business Press.

Chung, C. J., & Wee, H. M. (2011). Short life-cycle deteriorating product remanufacturing in a green supply chain inventory control system. *International Journal of Production Economics*, *129*(1), 195–203. doi:10.1016/j.ijpe.2010.09.033

Clark, R. C., & Mayer, R. E. (2011). E-Learning and the Science of Instruction, Proven Guidelines for Consumers and Designers of Multimedia Learning (3rd ed.). Pfeiffer: San Francisco.

Cognizant. (2016). *People — Not Just Machines — Will Power Digital Innovation*. Retrieved from https://www.cognizant.com/whitepapers/People-Not-Just-Machines-Will-Power-Digital-Innovation-codex1850.pdf

Cognizant. (2017). *People — Not Just Machines — Will Power Digital Innovation*. Retrieved March 31, 2017 from https://www.cognizant.com/whitepapers/People-Not-Just-Machines-Will-Power-Digital-Innovation-codex1850.pdf

CONDUSEF. (2013). *Common mistakes of the SME entrepreneur*. Retrieved from http://www.condusef.gob.mx/index.php/empresario-pyme-como-usuarios-de-servicios-financieros/548-errores-mas-comunes-del-empresario-pyme

Consultores, A. (2016). *Pareto Diagram - Quality Tools*. Retrieved from http://www.aiteco.com/diagrama-de-pareto

Corsinovi, P., & Gaeta, D. (2015). *Managing the Quality Wines beyond Policies and Business Strategies. Review Of Contemporary Business Research, 4(1)*. doi:10.15640/rcbr.v4n1a3

Cortes, J. (1999). Factors of leadership and competitiveness in a financial services company. *Revista ABANTE*, *2*(1), 71–96.

Crook, C. (2008). Web 2.0 technologies for learning: The current landscape – opportunities, challenges, and tensions. Nottingham: Becta: Leading next generation learning.

Cropley, A. J. (1979). *Lifelong education: A Stocktaking*. Hamburg: UNESCO Institute for Education.

Delgado, N., & Delgado, D. (2003). The leader and leadership: Reflections. *Interamerican Journal of Library Science*, *2*(26), 75–88.

Demiray, U. (2010). *E-Learning practices, Cases on challenges facing e-learning and national development: Institutional Studies and Practices (Vol. 1)*. Eskisehir, Turkey: Anadolu University.

Devedžić, V. (2006). Introduction to Web-Based Education. *Semantic Web and Education*, 1 - 28. eEurope (2005). *E-learning*. Retrieved January 12, 2017 from http://europa.eu.int/information_society/eeurope/2005/all_about/elearning/index_en.htmhttp://www.elearningconference.org

Diario Oficial de la Federación de México. (2013). *Decree amending and adding various provisions of the Political Constitution of the United Mexican States, in Energy Matters*. Retrieved from http://www.dof.gob.mx/nota_detalle.php?codigo=5327463&fecha=20/12/2013

Dickson, S. (1980). Personality and the Effective Teacher. *Journal of Teacher Education*, *12*(3), 335–337.

Digital Transformation. (2017). *A Roadmap For Billion-Dollar Organizations*. Retrieved March 31, 2017 from https://www.capgemini.com/resource-file-access/resource/pdf/Digital_Transformation__A_Road-Map_for_Billion-Dollar_Organizations.pdf

Dobre, I. (2010). Studiu critic al actualelorsisteme de e-learning, Academia Română. Institutul de cercetăripentruinteligenţăartificială: Bucureşti.

Duffy, J.E. (2014). Office Suite. *PC Magazine*.

Ellison, N. B., & Boyd, D. (2013). Sociality through Social Network Sites. In W. H. Dutton (Ed.), *The Oxford Handbook of Internet Studies* (pp. 151–172). Oxford: Oxford University Press.

Emelo, R. (2011). What if millennial ran your mentoring program. *Chief Learning Officer*, *10*(5), 32–36.

Entrepreneur, S. (2011). *How to do empowerment in your company*. Retrieved from http://www.soyentrepreneur.com/como-hacer-empowerment-en-tu-empresa.htm

Ertl, B. (2010). *E-Collaborative knowledge construction: Learning from computer supported and virtual environments*. Retrieved from http://www.library.uiuc.edu/proxy/go.php?url=http://services.igiglobal.com/resolvedoi/resolve.aspx?doi=10.4018/978-1-61520-729-9

Ethos BeathChapman. (2016, July 19). *The liquid workforce: A well of innovation*. Retrieved from http://ethosbc.com/industryinsights/theliquidworkforceawellofinnovation

EY. (2017). *The digitization of everything How organizations must adapt to changing consumer behavior*. Retrieved 31 March 2017, from http://www.ey.com/Publication/vwLUAssets/The_digitisation_of_everything_-_How_organisations_must_adapt_to_changing_consumer_behaviour/$FILE/EY_Digitisation_of_everything.pdf

FastTrack Press Room. (2016, July 14). *How can you facilitate the liquid workforce?* Retrieved from https://www.fasttrack.com.au/blog/howcanyoufacilitatetheliquidworkforce

Feldman, D. (1989). Careers in Organizations: Recent Trends and Future Directions. *Journal of Management*, *15*(2), 135–156. doi:10.1177/014920638901500202

Ferrimen, J. (2013). *Three types of e-learning*. Retrieved January 18, 2017 from https://www.learndash.com/3-types-of-elearning

Forrestera, S. V., Ustinovaa, G. H., Kosyakovaa, I. V., Ronzhinab, N. V., & Suraevac, M. O. (2016). Human Capital in the Innovative Conditions. *Mathematics Education*, *11*(8), 3048–3065.

Friedman, M. (1970, September 13). The Social Responsibility of Business is to increase profits. *The New York Times Magazine.*

García, M., Posada, V., & Hernández, C. (2012). *Motivation and reward systems and their impact on production, Contributions to the Economy.* Retrieved from http://www.eumed.net/ce/2012/gvr.html

García, J. (2013). *Diagnostic study to determine the leadership profile of the manager and his collaborators in the operational management, according to the full range model of Bass and Avolio: Case of a chemical industrial company in Coatzacoalcos.* Veracruz, México: Revista Caribeña de Ciencias Sociales.

G-Cube. (n. d.). *Different types of e-learning.* Retrieved January 18, 2017 from http://www.gc-solutions.net/resources/articles/different-types-of-e-learning-and-what-suits-me-best.html

Gilbert, J. (2011). The Millennial: A new generation of employees, a new set of engagement policies. *Ivey Business Journal, 75*(5), 26–28.

Globescan. (2013). *2.5 Billion Aspirational Consumers Mark Shift in Sustainable Consumption.*

Gloor, P. A., Paasivaara, M., Schoder, D., & Willems, P. (2008). Finding collaborative innovation networks through correlating performance with social network structure. *International Journal of Production Research, 46*(5), 1357–1371. doi:10.1080/00207540701224582

Good, C. V. (1973). *Dictionary of Education.* New York: McGraw-Hill Book Company.

Goodwill. (2016). *Gig Economy Provides New Options to "Liquid Workforce".* Retrieved from http://www.goodwill.org/blog/newsupdates/gigeconomyprovidesnewoptionstoliquidworkforce

Gopal, B. V., & Anandan, K. (2013). Attitude towards e-Learning in Classroom Instruction among the B. Ed. Students at Colleges of Education. *Seminar on Current perspectives on Education.* Retrieved from http://www.languageinindia.com/feb2013/gopalanandan.pdf

Grohowski, R., McGoff, C., Vogel, D., Martz, B., & Nunamaker, J. (1990). Implementing electronic meeting systems at IBM: Lessons learned and success factors. *Management Information Systems Quarterly, 14*(4), 369–383. doi:10.2307/249785

Guiney, P. (2015). E-learning in the Workplace- An annotated bibliography. New Zealand: Ministry of Education. Retrieved from www.educationcounts.govt.nz

Halder, B. (2014). Evolution of crowd sourcing: Potential data protection, privacy and security concerns under the new media age. *Revista Democracia Digital e Governo Eletrônico*, *1*(10), 377–393.

Harasim, L. (2006). A history of e-learning: shift happened. In J. Weiss, J. Nolan, J. Hunsinger, & P. Trifonas (Eds.), *The international handbook of virtual learning environments*. Dordrecht: Springer. doi:10.1007/978-1-4020-3803-7_2

Hava, H. T., & Erturgut, R. (2010). An evaluation of education relations together with technology, employment and economic development components. *Procedia: Social and Behavioral Sciences*, *2*(2), 1771–1775. doi:10.1016/j.sbspro.2010.03.982

Hedge, N., & Hayward, L. (2004). Redefining roles, University E-learning contributing to Life-long learning in a networked world. *E-learning*, *1*(1), 128–145.

Helbing, D. (2013). Globally networked risks and how to respond. *Nature*, *497*(7447), 51–59. doi:10.1038/nature12047 PMID:23636396

Hinchcliffe, D. (2017). The new digital workplace: How enterprises are preparing for the future of work. *ZDNet*. Retrieved March 31, 2017 from http://www.zdnet.com/article/the-new-digital-workplace-how-enterprises-are-preparing-for-the-future-of-work

Hoang, M. H. (2015). New models of human resource management in China and India. *Asia Pacific Business Review*, *21*(4), 560–562. doi:10.1080/13602381.2014.976976

Hoegg, J., Alba, J. W., & Dahl, D. W. (2010). The good, the bad, and the ugly: Influence of aesthetics on product feature judgments. *Journal of Consumer Psychology*, *20*(4), 419–430. doi:10.1016/j.jcps.2010.07.002

Hrastinski, S. (2008). Asynchronous & synchronous e-learning: A study of asynchronous and synchronous methods. *EDUCAUSE Quarterly*, *31*(4), 51–55.

Hughes, J. (2010). But it's not just developing like a learner, it's developing as a person – Reflections on e-portfolio-based learning. In R. Sharpe, H. Beetham, & S. De Freitas (Eds.), *Rethinking Learning for a Digital Age: How Learners Are Shaping Their Own Experiences*. New York: Routledge.

Humbert, M. (2007). *Technology and Workforce: Comparison between the Information Revolution and the Industrial Revolution*. Berkeley: University of California.

Impact Guru. (n. d.). Retrieved January 3, 2017, from https://www.impactguru.com

INEGI. (2010). Mexico en cifras. Retrieved from http://www.inegi.org.mx/Censos(2010)

INEGI. (2015). Business life expectancy. *Boletín de prensa*, *87*(15), 1 - 2.

Instituto nacional de seguridad e higiene en el trabajo. (1998). NTP 213: Job satisfaction: evaluation survey. Retrieved from http://www.insht.es/InshtWeb/Contenidos/Documentacion/FichasTecnicas/NTP/Ficheros/201a300/ntp_213.pdf

Interian, J. (2016). Up in the air: Harmonizing economy through airbnb regulations. *Boston College International and Comparative Law Review*, 39, 129–161.

Investopedia (n.d). *Human capital*. Retrieved December 10, 2016 from http://www.investopedia.com/terms/h/humancapital.asp

Investopedia Staff. (2011, January 19). Virtual Assistant. Retrieved July 21, 2017 from http://www.investopedia.com/terms/v/virtual-assistant.asp

ISACA. (2013). *CGEIT Review Manual 2013*. Illinois: ISACA Publishing.

Jawadekar, W. S. (2006). *Management Information Systems, Text and cases* (3rd ed.). New Delhi: Tata McGraw-Hill Publishing Company Limited.

Jayme. (2011). The University of Colorado. Social Networking: The Unharnessed Educational Tool. *Undergraduate Research Journal*, *4*(1).

Jefferson, O. (n. d.). Workplace Equality in International Organizations: A Way Forward. *SSRN Electronic Journal*. doi:10.2139/ssrn.1236602

Jhaveri, H., & Choski, A. (2014). Crowd funding at India: a study of Indian online crowd funding platforms. Retrieved 30 January, 2017 from https://www.researchgate.net/publication/272984988_CROWDFUNDING_AT_INDIA_A_STUDY_OF_INDIAN_ONLINE_CROWDFUNDING_PLATFORMS

Kamath, N. (2012). *Personal Data Privacy in the Online Context. In Computers Internet & E-Commerce, a Guide to cyber laws & The Information Technology Act* (5th ed., p. 314). Delhi: Universal Law Publishing.

Kaplan, R., & Norton, D. (2005). The Balance Scorecard: Measures that Drive Performance. *Harvard Business Review America Latina*, *83*(7), 102–110.

Kavanagh, M. J., Thite, M., & Johnson, R. D. (2015). Human resource information systems: Basics, applications and future directions (3rd ed.). New York: Sage Publishers.

Ketto. (n. d.). Retrieved January 3, 2017, from www.ketto.org/support/faqs.php

Khan, A. (1996). Utilization of communication technologies for distance. In K. Murali Manohar (Ed.), Distance education theory and practice: Media and Communication Technology.

Khan, B. H. (2001). A framework for web-based learning. *Web-based training*, 75 - 98.

Khan, K. U., & Badii, A. (2012). Impact of E-learning on higher education: Development of an e-learning framework. *Life Science Journal*, *9*(4), 4073–4082.

Kisanga, D. H., & Ireson, G. (2014). Challenges and strategies on adoption of e-learning in Tanzanian higher learning institutions: lessons to future adopters. In *Proceedings of the International Conference on Information Communication Technologies in Education ICICTE '14*, Kos, Greece (pp. 5 - 8).

Kramer, L. W. (2010). Generational diversity. *Dimensions of Critical Care Nursing*, *29*(3), 125–128. doi:10.1097/DCC.0b013e3181d24ba9 PMID:20395729

Kucharčíková, A. (2011). Human Capital – definitions and approaches. *Human Resources Management & Ergonomics*, *5*(2), 60–70.

Kumar, V., Zhang, X., & Luo, A. (2014). Modeling customer opt-in and opt-out in a permission-based marketing context. *JMR, Journal of Marketing Research*, *51*(4), 403–419. doi:10.1509/jmr.13.0169

Kyndt, E., Raes, E., Dochy, F., & Janssens, E. (2012). Approaches to learning at work: Investigating work motivation, perceived workload, and choice independence. *Journal of Career Development*, *40*(4), 271–291. doi:10.1177/0894845312450776

La Rosa, M. (2016). Interview with Michael Rosemann on The Role of Business Process Management in Modern Organizations. *Business & Information Systems Engineering*, *58*(1), 89–91. doi:10.1007/s12599-015-0419-8

LaPrade, K., Gilpatrick, M., & Perkins, D. (2014). Impact of Reflective Practice on Online Teaching Performance in Higher Education. *MERLOT Journal of Online Learning and Teaching*, *10*(4). Retrieved from http://jolt.merlot.org/vol10no4/Gilpatrick_1214.pdf

Lau, R. W., Yen, N. Y., Li, F., & Wah, B. (2014). Recent development in multimedia e-learning technologies. *World Wide Web (Bussum)*, *17*(2), 189–198. doi:10.1007/s11280-013-0206-8

LDSA (Learning Skills Development Agency). (2015). *New trends in education and learning - online learning*. Retrieved January 12, 2017 from http://www.lsda.org.uk/new-trends-in-education-and-learning-online-learning.html

Lee Rainie, S. K. (2013). *Anonymity, Privacy, and Security Online*. Retrieved 2016 from http://www.pewinternet.org/2013/09/05/anonymity-privacy-and-security-online

Lee, J. (2004). The Effects of Family Ownership and Management on Firm Performance. *SAM Advanced Management Journal*, *69*(4), 46–53.

Lerer, L., & Piper, M. (2003). *Digital strategies in the pharmaceutical industry*. Houndmills, Basingstoke, Hampshire: Palgrave Macmillan. doi:10.1057/9780230598799

Liebowitz, J., & Frank, M. (2011). *Knowledge Management and E-Learning*. U.S.A.: Taylor & Francis Group.

López, J. (2005). Labor Motivation and Human Resource Management and the Theory of Frederick Herzberg. *Gestión en el Tercer Milenio*, *8*(15), 25–36.

Lotysz, S., & Helerea, E. (2014). Programme and abstracts: technology in times of transition. In *Proceedings of the 41st ICOHTEC symposium,* Transilvania University of Brasov, Brasov, Rumanía.

Luck, L. T., Hashim, F., & Din, S. Z. M. (2012). A Creative and Literary Writing Digital LMS in Supporting Writers Group Learning and Knowledge Sharing among Creative Writers. *Procedia: Social and Behavioral Sciences*, *67*, 238–249. doi:10.1016/j.sbspro.2012.11.326

Lytras, M. D., Mathkour, H. I., Abdalla, H., Al-Halabi, W., Yanez-Marquez, C., & Siqueira, S. W. M. (2015). An emerging–Social and emerging computing enabled philosophical paradigm for collaborative learning systems: Toward high effective next generation learning systems for the knowledge society. *Computers in Human Behavior*, *51*, 557–561. doi:10.1016/j.chb.2015.06.004

Mann, N. (2015, October 26). How Does a POS System Work? Retrieved July 21, 2017 from http://www.businessbee.com/resources/profitability/pos-system-work

Marc Lieberstein, A. T. (2012). *Crowd sourcing and Intellectual Property Issues*. Retrieved from http://www.acc.com/legalresources/quickcounsel/caipi.cfm

Mariani, A., Annunziata, A., Nacchia, F., & Vastola, A. (2016). Crowd sourcing in Wine Business: Co-Creation and Fundraising Experiences. In Successful Social Media and Ecommerce Strategies in the Wine Industry (pp. 93 - 114). Palgrave Macmillan US.

Maslow, A. (1943). A theory of motivation. *Psychological Review*, *50*(4), 370–396. doi:10.1037/h0054346

McFarland, W., & Jestaz, D. (2016). Talent development in the digital age: A neuroscience perspective. *Industrial and Commercial Training*, *48*(2), 74–79. doi:10.1108/ICT-09-2015-0062

McGill, T. J., Klobas, J. E., & Renzi, S. (2014). Critical success factors for the continuation of e-learning initiatives. *The Internet and Higher Education*, *22*, 24–36. doi:10.1016/j.iheduc.2014.04.001

McGuigan, E. (2016). *Liquid workforce: building the workforce for today's digital demands.* Telegraph. Retrieved from http://www.telegraph.co.uk/sponsored/business/digitalleaders/futureseries/12161816/liquidworkforce.html

McNely, B.J. (2009). Backchannel Persistence and Collaborative Meaning. In *Proceedings of SIGDOC'09*, October 5 – 7.

McPherson, M., & Nunes, M. (2006). Organisational issues for e-learning critical success factors as identified by HE practitioners. *International Journal of Educational Management*, *20*(7), 542–558. doi:10.1108/09513540610704645

Mehmood, S., & Taswir, T. (2013). The Effects of Social Networking Sites on the Academic Performance of Students in College of Applied Sciences. *International Journal of Arts and Commerce*, *2*, 113.

Meyerson, B. (2012). *IBM News Room.* Retrieved from http://www-03.ibm.com/press/us/en/pressrelease/39685.wss

Microsoft. (2013). *Internet of things - GfK.* Retrieved November 2013 from https://www.gfk.com/fileadmin/.../Internet_of_things_-_whitepaper_-_digitaal.pdf

Milaap. (n. d.). Retrieved January 3, 2017, from https://milaap.org

Mollick, E. (2013). Step away from the crowd? Crowd funding, Venture Capital and the Selection of Entrepreneurs. Retrieved 29 January, 2017 from ssrn.com/abstract=2239204

Mollick, E. (2013). The Dynamics of Crowd funding: An Exploratory Study. *Journal of Business Venturing*, *29*(1), 1–16. doi:10.1016/j.jbusvent.2013.06.005

Moore, J. L., Dickson-deane, C., & Gaylen, K. (2015). e-Learning, online–learning, and distance learning environments: Are they the same? *The Internet and Higher Education*, *14*(1), 129–135.

Motschnig-Pitrik, R., & Standl, B. (2012). *Person-centered technology enhanced learning: Dimensions of added value. Computers in Human Behaviour*. doi:10.1016/j.chb

Nagarajan, P., & Jiji, G. W. (2010). Online educational system (e-learning). *International Journal of u - and e - Service. Science and Technology*, *3*(4), 37–48.

Naidoo, V. (2017). E-Learning and Management Education at African Universities. In N. Baporikar (Ed.), *Management Education for Global Leadership* (pp. 181–201). Hershey, USA: IGI Global. doi:10.4018/978-1-5225-1013-0.ch009

Nair, A., & Gaur, V. (2016, September 10). *ICICI Bank introduces software robotics*. Retrieved from: http://economictimes.indiatimes.com/industry/banking/finance/banking/icici-bank-introduces-software-robotics/articleshow/54258671.cms?utm_source=facebook.com&utm_medium=referral&utm_campaign=ETFBMain

Nelson, B. (1996). 1001 ways to reward employees. Colombia: Norma.

Neuner, J. (2013). *40% of America's workforce will be freelancers by 2020*. Retrieved from http://qz.com/65279/40-of-americas-workforce-will-be-freelancers-by-2020

Nicoll, K., & Fejes, A. (2011). Lifelong learning: A Pacific Know How. *Studies in Philosophy and Education*, *30*(4), 403–417. doi:10.1007/s11217-011-9235-x

Nisar, T. M. (2002). Organisational Determinants of E-learning. *Industrial and Commercial Training*, *34*(7), 256–262. doi:10.1108/00197850210447246

Njenga, J. K., & Fourie, L. C. H. (2010). The myths about e-learning in higher education. *British Journal of Educational Technology*, *41*(2), 199–212. doi:10.1111/j.1467-8535.2008.00910.x

NorénCreutz, I., & Wiklund, M. (2014). Learning paradigms in workplace e-learning research. *Knowledge Management & E-Learning*, *6*(3), 299–315.

North, D. C. (1990). *Institutions, institutional Change, and economic performance*. New York: Norton. doi:10.1017/CBO9780511808678

North, D. C. (2005). *Understanding the Process of Economic Change*. Princeton, University Press. doi:10.1515/9781400829484

Oketch, M. O. (2006). Determinants of human capital formation and economic growth of African countries. *Economics of Education Review*, *25*(5), 554–564. doi:10.1016/j.econedurev.2005.07.003

Olaniyan, D. A., & Okemakinde, T. (2008). Human capital theory: Implications for human resource development. *Pakistan Journal of Social Sciences*, *5*(5), 479–483.

Omara, N. D., Hassana, H., & Atan, H. (2012). Student Engagement in Online Learning: Learners Attitude toward E-Mentoring. *Procedia-Social and Behavioural Sciences*, *67*. doi:10.1016/j.sbspro.2012.11.351

Ong, C. S., Lai, J. Y., & Wang, Y. S. (2004). Factors affecting engineers' acceptance of asynchronous e-learning systems in high-tech companies. *Information & Management, 41*(6), 795–804. doi:10.1016/j.im.2003.08.012

Orange Business Services. (2017). The Next Workplace Revolution– how technology is changing jobs. Retrieved March 31, 2017 from http://www.orange-business.com/en/blogs/connecting-technology/innovation/the-next-workplace-revolution-how-technology-is-changing-jobs

Organizational Design. (2017). *Digital Technologies and Trends 2017-How Organizations Manage Innovation-Organizational Design*. Retrieved 31 March 2017, from https://gernotkapteina.com/2017/01/17/digital-technologies-and-trends-how-organizations-become-digital

Overton, L. (2004). *Linking learning to business: research report 2004. Reading*. UK: Bizmedia.

Ozkan, S., & Koseler, R. (2009). Multi-dimensional student's evaluation of e-learning systems in the higher education context: *An empirical investigation. Computers & Education, 53*(4), 1285–1296. doi:10.1016/j.compedu.2009.06.011

Ozuorcun, N. C., & Tabak, F. (2012). Is m-learning versus e-learning or are they supporting each other? *Procedia: Social and Behavioral Sciences, 46*, 299–305. doi:10.1016/j.sbspro.2012.05.110

Pallud, J. (2016). Impact of interactive technologies on stimulating learning experiences in a museum. *Information & Management*.

Palmer, M. (2017). *8 Innovation Trends for Digital Insurance in 2016. The TIBCO Blog*. Retrieved March 31, 2017 from http://www.tibco.com/blog/2016/01/15/8-innovation-trends-for-digital-insurance-in-2016

Panetta, K. (2017). *Gartner's Top 10 Strategic Technology Trends for 2017-Smarter With Gartner*. Retrieved March 31, 2017 from http://www.gartner.com/smarterwithgartner/gartners-top-10-technology-trends-2017

Parkes, I. (2013). *Colemanparkes*. Retrieved 2016 from http://colemanparkes.com

Pathak, S., & Kaur, P. (2016). Crowd funding: an innovation in the age of digital marketing. International Journal for Innovations in Engineering Management and Technology, 1(1), 26 - 34.

Paulsen, M. F. (2003). Experiences with Learning Management Systems in 113 European Institutions. *Journal of Educational Technology & Society, 6*(4), 134–148.

Pegasystems. (2016). *Liquid Workforces Will Let Businesses Put Customers at the Center the digital age is forcing companies to be more fluid about organizational frameworks*. Retrieved from http://digitalevolution.eiu.com/liquidworkforceswillletbusinessesputcustomersatthecenter

Peiró. (2000). Psychology of the Organization. *Universidad Nacional de Educación a Distancia, Madrid*, *159*, 165–170.

Peng, M. W. (2000). *Business strategies in transition Economies* (pp. 42–44). CA: Sage. Pp.

Peng, M. W. (2012). *Estrategia global*. México: Cengage Learning.

Percival, J. C., Cozzarin, B. P., & Formaneck, S. D. (2013). Return on investment for workplace training: The Canadian experience. *International Journal of Training and Development*, *17*(1), 20–32. doi:10.1111/ijtd.12002

Petrou, A. (2011). Assessing Integration of Infrastructure 2.0 Tools in Open Learning Websites: Highlighting a Need for Purposeful Design. In Proceedings of Informing Science & IT Education Conference (pp. 427 - 441).

Phillips, J. (2010). Open innovation typology. *International Journal of Innovation Science*, *2*(4), 175–183. doi:10.1260/1757-2223.2.4.175

Power, B. (2014, April 23). How GE Applies Lean Startup Practices. *Harvard Business Review*. Retrieved from https://hbr.org/2014/04/how-ge-applies-lean-startup-practices

Pride, W. M., Huges, R. J., & Kapoor, J. R. (2017). *Foundations of Business*. Singapore: Cengage Learning.

PROMEXICO. (n. d.). *PYMES is a fundamental link for the progress of Mexico*. Retrieved from http://www.promexico.gob.mx/negocios-internacionales/pymes-eslabon-fundamental-para-el-crecimiento-en-mexico.html

PWC. (2011). Millennial at work: Reshaping the workplace. *Price Waterhouse Coopers*. Retrieved from https://www.pwc.com/m1/en/services/consulting/documents/millennials-at-work.pdf

Rani, P., & Katyal, H. (2015). Venture capital in India: sector - wise analysis. International Journal of Research & Development in Technology and Management Science, 21(6), 72 - 80.

Real Academia Española. (2017). *Diccionario de la lengua española*. Retrieved from http://dle.rae.es/?id=aBuhX28

Rick, T. (2017). *The real challenge of digitization is not technology*. Retrieved March 31, 2017 from https://www.torbenrick.eu/blog/technology/the-real-challenges-of-digitization-is-not-technology

Rifkin, J. (2011). *The Third Industrial Revolution*. Palgrave Macmillan.

Rifkin, J. (2014). *La sociedad de coste marginal cero: el internet de las cosas, el procomún colaborativo y el eclipse del capitalismo*. Barcelona: Paidós.

Rittgen, P. (2010, June). Success factors of e-collaboration in business process modeling. In *Proceedings of the International Conference on Advanced Information Systems Engineering* (pp. 24 - 37). doi:10.1007/978-3-642-13094-6_4

Rodríguez, D. (2006). *Organizational Diagnosis*. México: Alfa Omega.

Rosa, F. (2017). What do digital; automation and outsourcing have in common. *CIO*. Retrieved March 31, 2017 from http://www.cio.com/article/3020058/leadership-management/what-do-digital-automation-and-outsourcing-have-in-common.html

Rosende, M., & Miró, M. (2013). Recent trends in automatic dynamic leaching tests for assessing bioaccessible forms of trace elements in solid substrates. *Analytical Chemistry, 45, 67-78*. doi:10.1016/j.trac.2012.12.016

Rubenson, K. (2009). Lifelong learning: Between humanism and global capitalism. In Jarvis (Ed.) The Routledge International Handbook of Lifelong learning. Routledge: London & New York.

Rubenson, K. (2006). The Nordic model of lifelong learning. *Compare: A Journal of Comparative Education*, *36*(3), 327–341. doi:10.1080/03057920600872472

Russo, A., & Quaglieri, A. (2014). La lógica espacial del intercambio de casas: una aproximación a las nuevas geografías de lo cotidiano en el turismo contemporáneo. *Revista electrónica de georgrafía y ciencias sociales*.

Ryngaert, C. (2010). *The Immunity of International Organizations Before Domestic Courts: Recent trends. International Organizations Law Review, 7(1), 121-148*. doi:10.1163/157237310x523803

Salas, E., & Gelfand, M. J. (2013). Introduction to the Special Issue: Collaboration in multicultural environments. *Journal of Organizational Behavior*, *34*(6), 735–738. doi:10.1002/job.1880

Sambamurthy, V., Bharadwaj, A., & Grover, V. (2003). Shaping agility through digital options: Reconceptualizing the role of information technology in contemporary firms. *Management Information Systems Quarterly*, 27(2), 237–263.

Sánchez, E., & Rodríguez, A. (2010). 40 years of situational leadership theory: A review. *Revista Latinoamericana de Psicología, 42*(1), 29–39.

Sangrà, A., Vlachopoulos, D., & Cabrera, N. (2012). Building an inclusive definition of e-learning: An approach to the conceptual framework. *The International Review of Research in Open and Distributed Learning, 13*(2), 145–159. doi:10.19173/irrodl.v13i2.1161

Sarkar, A. (n. d.). Crowd funding in India: Issues and Challenges. Retrieved 29 January, 2017, from http://ssrn.com/abstract=2739008

Savitz, A. (2012). *The triple bottom line: How today's best-run companies are achieving economic, social and environmental success--and how you can too*. John Wiley & Sons.

Scagnoli, N. I., Buki, L. P., & Johnson, S. D. (2007). The Influence of Online Teaching on Face-To-Face Teaching Practices. *Journal of Asynchronous Learning Networks, 13*(2). Retrieved from http://files.eric.ed.gov/fulltext/EJ862351.pdf

Scientific American. (1855). Recent Foreign Inventions. *Scientific American, 11*(1), 3-3. doi:10.1038/scientificamerican09151855-3a

Shane, S. (2008). The Importance of Angel Investing in Financing the Growth of Entrepreneurial Ventures (Rep. No. 331).

Sharadha, D.G., & Parameswaram, E.G. (2008). Teacher Characteristics and Learning in the Classroom. *Edutracks, 11*(3).

Shea, R. H. (2002). E-Learning Today; As an industry shakes out, the survivors offer no frills education for grown-ups. *U.S. News & World Report, 133*(1), 54–56.

Shivaprasad, G., & Kannadas, S. (2016). Is Endowment of Crowd Funding a Supplant of Traditional Source? International Journal of Scientific Research, 5(2), 138 - 141.

Shukla, V. K., & Agrawal, P. (2013). Towards Semantical E-Learning from E-learning. *International Journal of Science and Research, 2*(11), 243–247.

Sidibé, M., & Campbell, J. (2015). Reversing a global health workforce crisis. *Bulletin of the World Health Organization, 93*(1), 3–3. doi:10.2471/BLT.14.151209 PMID:25558098

Siemens, G. (2014). *Connectivism: A learning theory for the digital age*. Retrieved January 24, 2017 from http://er.dut.ac.za/handle/123456789/69

Simões, T., Rodrigues, J. J., & De la Torre, I. (2013). Personal Learning Environment Box (PLEBOX): A new approach to E-learning platforms. *Computer Applications in Engineering Education*, *21*(S1), 100–109. doi:10.1002/cae.20537

Simons, M. (2016, April 22). *Transforming the Workforce in the Digital Era.* Retrieved from http://www.cio.co.uk/it-leadership/transforming-workforce-in-digital-era-3638972

Singh, H. (2003). Building effective blended learning programs. *Educational Technology-Saddle Brook Then Englewood Cliffs NJ*, *43*(6), 51–54.

Sloman, M., & Rolph, J. (2003). Change Agenda. *CIPD*. Retrieved 5 September 2009, from http://www.cipd.co.uk/NR/rdonlyres/2716D932-3FEF-4559-9667-2AD7006EA13F/0/elrn_curve.pdf

Sloman, M., & Rolph, J. (2003). *E-learning the learning curve: the change agenda. Institute of Personnel and Development.* London: IPD.

SmashFly Blog. (2017). *3 Trends Defining Hiring Success In The Digital Economy-SmashFly Blog*. Retrieved March 31, 2017 from http://blog.smashfly.com/2016/04/14/3-trends-defining-hiring-success-in-the-digital-economy

Sorokin, A., & Forsyth, D. (2008, June). Utility data annotation with Amazon mechanical Turk. In *Proceedings of the IEEE Computer Society Conference on* Computer Vision and Pattern Recognition Workshops CVPRW '08. IEEE.

Stamford, C. (2013). Gartner Says Potential Size and Diversity of the Internet of Things Mask Immediate Opportunities. *Gartner*. Retrieved 2016 from http://www.gartner.com/newsroom/id/2564916

Start51. (n. d.). Retrieved January 3, 2017, from http://www.start51.com/help

Sterling, C. H. (2009). *Encyclopedia of Journalism*. Thousand Oaks, CA: Sage Reference. doi:10.4135/9781412972048

Stollak Matthew, J., Vandenberg, A., Burklund, A., & Weiss, S. (2011). Getting Social: The Impact of Social Networking Usage on Grades among College Students. *Proceedings of ASBBS*, *18*, 859–860.

Streitfeld, D. (2014, October 15). Airbnb listing mostly illegal. *New York Times*.

Sukaini A. K. M., & Zhang, J. (2016). Crowd sourcing and e-Commerce. Chinese Online Reviews.

Sung, S., & Choi, J. (2014). Do organizations spend wisely on employees? Effects of training and development investments on learning and innovation in organizations. *Journal of Organizational Behavior*, *35*(1), 393–412. doi:10.1002/job.1897 PMID:25598576

Tannenbaum, R., & Schmidt, W. H. (1973). How to choose a leadership pattern. *Harvard Business Review*, *51*(3), 162–173.

Tavangarian, D., Leypold, M. E., Nölting, K., Röser, M., & Voigt, D. (2004). Is e-learning the Solution for Individual Learning. *Electronic Journal of E-learning*, *2*(2), 273–280.

Taylor, K. (2013). *6 Top Crowd funding Websites: Which One Is Right for Your Project?* Retrieved from http://www.forbes.com/sites/katetaylor/2013/08/06/6-top-crowdfunding-websites-which-one-is-right-for-your-project/#668be67265f8

Toor, K. K. (2014). A Study of Teacher Effectiveness, General Intelligence and Creativity of Secondary School Teachers. *MIER Journal of Educational Studies, Trends & Practices*, *4*(1).

Uber. (2016). Retrieved from https://www.uber.com

Urbado, D., Díaz, J., & Hernández, R. (2007). Evolución y principios de la teoría económica institucional. Una propuesta de aplicación para el análisis de los factores condicionantes de la creación de empresas. *Investigaciones Europeas de Dirección y Economía de la Empresa*, *13*(3), 183–198.

Veblen, T. (1965). *La teoría der la empresa de negocios (1904)*. Buenos Aires: Eudeba.

Venture-Capital-Investment.co.uk. (n. d.). Venture capital investment vs. angel investment. Retrieved January 3, 2017, from http://www.venture-capital-investment.co.uk/venture-capital-investment-vs-angel-investment.php

Victor, S. R. (2013). Teacher-Trainees Attitude towards ICT. *Journal of Education and Practice*, *4*(19).

Vijayalakshmi, A. (2005). Teacher Effectiveness and Job Satisfaction of Women Teachers. *Edutracks*, *4*(7).

Villarreal, O. (2005). La internacionalización de la empresa y la empresa multinacional: Una revisión conceptual contemporánea. *Cuadernos de Gestión*, *5*(2), 55–73.

Von Ahn, L., Ginosar, S., Kedia, M., & Blum, M. (2007). Improving image search with phetch. In *Proceedings of the IEEE international conference on Acoustics, speech and signal processing ICASSP '07* (Vol. 4, pp. IV–1209). IEEE.

Wagner, N. L., Hassanein, K., & Head, M. M. (2008). Who is Responsible for E-Learning Success in Higher Education? A Stakeholders' Analysis. *Journal of Educational Technology & Society*, *11*(3), 26–36.

Wells, D. (2016, May 9). If you're not tapping into the Liquid Workforce, You're Not Innovating. *Inc.com*. Retrieved from http://www.inc.com/donna-wells/if-you-re-not-tapping-into-the-liquid-workforce-you-re-not-innovating.html

Welsh, E. T., Wanberg, C. R., Brown, K. G., & Simmering, M. J. (2003). E-learning: Emerging uses, empirical results and future directions. *International Journal of Training and Development*, *7*(4), 245–258. doi:10.1046/j.1360-3736.2003.00184.x

Wen, I. (2009). Factors affecting the online travel buying decision: a review. *International Journal of Contemporary Hospitality Management*, *21*(6), 752-765.

Wilson, J. (2000, February). Lessons of a Virtual Timetable: Education. *The Economist*.

Wishberry. (n. d.). Campaign Management Fees. Retrieved January 3, 2017, from https://www.wishberry.in/pricing/#/pricing

Wishberry. (n. d.). Retrieved January 3, 2017, from https://www.wishberry.in/rules

Woo, Y., & Reeves, T. C. (2007). Meaningful interaction in web-based learning: A social constructivist interpretation. *The Internet and Higher Education*, *10*(1), 15–25. doi:10.1016/j.iheduc.2006.10.005

Wright, A. W., Hamilton, B., Mighty, J., Scharfe, E., Muirhead, B., & Vail, S. (2014). Shared Modular Course Development: A Feasibility Study. *Centre for Teaching and Learning Reports*. Retrieved January 22, 2017, from http://scholar.uwindsor.ca/ctlreports/1

Xiberta, P., & Boada, I. (2016). A new e-learning platform for radiology education (RadEd). *Computer Methods and Programs in Biomedicine*, *126*, 63–75. doi:10.1016/j.cmpb.2015.12.022 PMID:26774237

Ya Ni, A. (2013). Comparing the Effectiveness of Classroom and Online Learning: Teaching Research Methods. *Journal of Public Affairs Education*, *19*(2), 199–215. Retrieved from http://www.naspaa.org/JPAEMessenger/Article/VOL19-2/03_Ni.pdf

Yukl, G. H. (1994). *Leadership in Organizations* (3th ed.). Englewood Cliffs: Prentice - Hall.

Zaino, G. (2005). Freelance Workers: Individual and Corporate Perspectives. *Stcwdc.org*. Retrieved from http://wdcb.stcwdc.org/wp-content/PDF/Individual_and_Corporate_Perspectives.pdf

Zeithamal, V. A., Parasuraman, A., & Malhotra, A. (2002). Service quality delivery through web sites: A critical review of extant knowledge. *Journal of the Academy of Marketing Science*, *30*(4), 362–375. doi:10.1177/009207002236911

Zhang, D., Zhao, J. L., Zhou, L., & Nunamaker, J. F. Jr. (2004). Can e-learning replace classroom learning? *Communications of the ACM*, *45*(5), 75–79. doi:10.1145/986213.986216

About the Contributors

Punita Duhan is a faculty in Business Administration with Department of Training and Technical Education, Govt. of NCT of Delhi for the last 18 years and currently pursuing research in Social media from Institute of Management Studies, Banaras Hindu University, Varanasi. Her research interests include Consumer behaviour, Behavioural Marketing, Social and Digital technologies, Green Marketing, Digital Marketing, e-Commerce, and m-Commerce. Prior to this, she has worked as Faculty of Management with University Business School, Punjab University, Chandigarh and Institute of Management and Technology, Faridabad. She has attended more than 35 national as well as international seminars/conferences and has presented papers in more than 25 national and international seminars and conferences. She has published more than 12 research papers with prestigious national and international publishers such as Sage, Palgrave, Springer, IGI global, Bloomsbury etc. She has published an edited book titled "Managing Public Relations and Brand Image through Social Media" with IGI Global. Another book is in pipeline with Apple Academic Press, Taylor & Francis Group. She is also on editorial board of peer reviewed prestigious journals. Besides, she has been actively involved in curriculum development and e-content development of various universities, National Institute of open schooling, Central Board of Secondary Education and Technical education departments of Delhi and Haryana state in management area. She has attended more than 30 training programs and workshops and has organized around 10 seminars/ conferences/ workshops.

Komal Singh is a visiting faculty in Business Administration with Department of Training and Technical Education, Govt. of NCT of Delhi from last 3 years. She has done Ph. D. in Human Resource Management from Mody University of Science and Technology, Rajasthan. Prior to this, she has a vast & diverse research experience of around 6 years in the field of Human Resources.

Rahul Verma is a visiting faculty in Management with Department of Training and Technical Education, Govt. of NCT of Delhi for the last 5 years. He is pursuing his Ph.D. in commerce and has attended more than 5 national as well as international seminars / conferences and has presented papers in more than 3 national and international seminars and conferences. He has to his credit more than 3 research papers in national and international journals. Prior to this, he has worked as Faculty of Management with various colleges under Guru Gobind Singh Indraprastha Univerity, Delhi, India.

* * *

Sulaiman Olusegun Atiku is a Postdoctoral Research Fellow in the School of Management, IT and Governance at the University of KwaZulu-Natal. He is a pragmatic researcher specialising on contemporary issues in Human Resource Management, Industrial Relations, Global Leadership Orientations and Practice, Modernisation and Environmental Sustainability, and Organisational Behaviour. He has over 9 years' experience in Higher Education. A native of Lagos, Sulaiman graduated from the University of KwaZulu-Natal with a PhD degree in Human Resource Management. The author acquired his Masters of Science degree in Human Resource Management and Industrial Relations at the Lagos State University, Nigeria. His Bachelor of Science (Honours) degree was also awarded at the Lagos State University in the field of Industrial Relations and Personnel Management. He has lectured several courses in his field home/abroad and published many scholarly articles in international journals. He is a member of the International Labour and Employment Relations Association (ILERA), Nigerian Institute of Management (NIM), Certified Institute of Warehousing and Materials Management (CIWM), and Institute of People Management (IPM) South Africa.

Indranil Bose, Ph.D. in Strategic Industrial Relations, is currently heading the School of Business, University of Bolton, UAE academic centre as the Dean. He is a seasoned academician and is having an overall teaching experience of around 17 years in India and UAE. As a passionate researcher, he has worked with many independent research projects and few funded projects on the areas of Industrial Relations, Strategic Human Resource Management, Green HR, HR scorecard, etc. He has presented many research papers in the international and national conferences in India, UAE, Oman and Malaysia. Many of his research papers have been published with reputed national and international journals, published from India, UAE, Singapore, Oman, Russia and UK. He is also a regular corporate trainer in India and UAE. He has also been conferred on several prestigious national level teaching and academic awards from India. He is the professional member of Society for Human Resource Management (USA), International Association for Advancement of Economic Growth, Hongkong, etc.

Anchal Chhabra is a visiting faculty in Management with Department of Training and Technical Education, Govt. of NCT of Delhi. She is an academician with a research aptitude. She has a keen eye for detail towards the work assigned and believes in constant up gradation to achieve professional results. A multifaceted personality, she holds BBA and MBA degree from GGSIPU. She is also the degree holder of M.Phil in Management and pursuing her doctorate in the same field to further strengthen her research acumen. She has participated and presented papers in various national and international conferences. She is NCFM certified in two modules from NSE. She has completed certification in Service Marketing from MOOC. In her rich work experience in academics, she has actively participated in various activities and worked towards the up gradation of research and learning.

Surbhi Gosain is a student of Management of Business Administration from Bhai Paramanand Institute of Business Studies (Govt. of NCT of Delhi) affiliated under Guru Gobind Singh Indraprastha University. Her area of specialization is finance. She had studied B.com (Hons) from University of Delhi. She had been a scholastic student during her school. She had passed her management exams with distinction. She is also a student of CA-IPCC and had undergone information technology training as well. She had also won various academic and non-academic certificates during her academic tenure. Her area of interest for research is mainly finance and marketing. she had conducted various research in the same field during her academic tenure. Having a commerce background, author is very much inclined towards the new age financial innovations and crowdfunding is one such area.

Manish Gupta is an Assistant Professor in the Department of Human Resource at IBS Hyderabad, IFHE University. His PhD thesis is in the area of 'work engagement'. Manish is a recipient of Junior Research Fellowship and National Eligibility Test certificate awarded by University Grants Commission, Government of India. He is Editorial Team Member of Australasian Journal of Information Systems. He is a member of ISDS, AIMA, DSI, and IBSAF. Currently, he is authoring books for publication in Oxford and Sage. He is editing two books with IGI Global, USA, and Apple Academic Press Inc., a member of the Taylor & Francis Group, USA. He is also editing three Special Issues for Advances in Developing Human Resources (Sage), Journal of Global Operations and Strategic Sourcing (Emerald), and industrial and Commercial Training (Emerald). His teaching areas include Human Resource Management, Entrepreneurship, and Organizational Behavior, among others. His scholastic work is available at https://scholar.google.co.in/citations?user=yJtwSkwAAAAJ&hl=en.

Vishwas Gupta is working as an assistant professor in Mittal School of Business, Lovely Professional University, Jalandhar. A PhD Scholar of Aligarh Muslim University, he has a very rich academic experience of 10 years. He has worked with few corporate companies for six years. His research areas are consumer psychology and retail marketing. He is author of two books on Marketing and Business Research Methodology. He has written several research papers and articles in various peer reviewed journals. He has been awarded with several prestigious academic awards like Best researcher award from Lovely Professional University in 2014, and Best faculty award from ICFAI National College in 2009. He is also professional member of reputed national and international bodies such as All India Management Association, India; Management Teachers Consortium (MTC), India.

Manisha Mathur, a heterogeneous personality, has finest revelations in her brilliant academic professional career. She has done her post graduates in two subjects one is Psychology (Industrial Relationship and Organizational behavior) and second is Information Technology. She did various Computer courses from NIIT Jodhpur. She is doing her research on "Cyber Crimes" as a research scholar of Mewar University, Chittorgarh, Rajasthan. Presently, she is associated as an Information Assistant with National Law University, Jodhpur. She has delivered her lectures in various institutions. She has keen interest in research in various fields and also interested to do social work.

Vannie Naidoo is a full-time staff member at the University of KwaZulu-Natal, South Africa, in the Faculty of Management and Law. Her field of teaching is in management, corporate strategy, project management and entrepreneurship. The research areas pursued by her are in the areas of management, marketing, service marketing, entrepreneurship and social media and e-learning or learning with new technologies. She has presented papers at international conferences throughout the world and has written on various issues in management and marketing in journal articles and books.

Lakshay Panchal is pursuing B.E. in Electronics and Communication Engineering (ECE) from Netaji Subhas Institute of Technology (NSIT), Dwarka, University Of Delhi. He has been honored with many scholarships and awards at school and college level. Also, since 2005, he is an active member of NGO - Leaders for Tomorrow.

Swati Panchal treats everyday experiences as a part of continuous learning process, pursuing adaptability, developing proficiency and acumen in dealing with situations and creating exceptional value in this dynamic profession exploring my skills to provide Quality Education. Presently Working as Sr. Coordinator, HOD, PGT Computer Science, IT Head at Mayo International School. Attended many in-service training for CCE, Python, etc. CCE training by Ratna Sagar. Won many awards like Board Result appreciation, 100% attendance award. Published books with Oswaal Publication are Sample Paper Informatics Practices Class XI(CBSE), Question Bank Of Informatics Practices class XI(CBSE), Sample Paper Informatics Practices Class XII(CBSE), Question Bank Of Informatics Practices class XII(CBSE), Sample Paper COMPUTER SCIENCE Class XII(CBSE), Question Bank Of COMPUTER SCIENCE class XII(CBSE), Question Bank of Computer Science class XII(ISC); Under Process, New IT Series of class 3, 4, 5 with Saraswati Publication House (ongoing project).

Ángel Daniel Rodriguez Ortega is Ph.D. in Administration Sciences from the UCC Business School, a Master's Degree in Computer Science from the Arturo Rosenblueth Foundation (Honorable Mention) and a degree in Computer Science from the Universidad Veracruzana. His training has been completed with IT certification programs and is currently a college professor at several universities in the city of Veracruz México. He does consulting projects and research in the areas of Information Systems, E - Commerce and Organizational Development.

Fabián Rojas Álvarez is research assistant at University Center for Economic and Managerial Sciences, University of Guadalajara. He holds a bachelor degree in Economics and a Master degree in Business and Economic Studies.

Neetu Singh has been working as Assistant Professor in Department of Pedagogical Sciences Dayalbagh Educational Institute (Deemed University) Agra, India. She has participated in various National and International conferences and has a large number of publications having good impact factor with National and International repute. She is working as Section Editor in "Second Edition of Handbook of Mobile Teaching and Learning", Publisher Springer USA. She is Co - Editor of Pahal Horizon: An International Biannual Research Journal in Social Sciences, Humanities and Management. She is member of editorial board of "Handbook of Research on Promoting Global Peace and Civic Engagement through Education" IGI Global Publication, USA. She is life member of various research organizations. She has a good academic record as B.SC., B.Ed., M.Ed., M.A. (Economics), M.A. (Education) in first division, Ph.D. Her research interests include ICT based Education, Innovative Learning Pedagogy, Special Education and Women Education.

José G. Vargas-Hernández is a member of the National System of Researchers of Mexico and a research professor at University Center for Economic and Managerial Sciences, University of Guadalajara. He has a Ph.D. in Public Administration and a Ph.D. in Organizational Economics. He has undertaken studies in Organizational Behavior and has a Master of Business Administration, published four books and more than 200 papers in international journals and reviews (some translated to English, French, German, Portuguese, Farsi, Chinese, etc.) and more than 300 essays in national journals and reviews. He has obtained several international Awards and recognition too.

Index

T

U

V

W

www.ingramcontent.com/pod-product-compliance
Lightning Source LLC
LaVergne TN
LVHW081331300926
849161LV00008B/1946
* 9 7 8 1 5 2 2 5 3 1 9 1 3 *